D0098688

HOW ANIMALS GRIEVE

How Animals Grieve

BARBARA J. KING

THE UNIVERSITY OF CHICAGO PRESS : CHICAGO AND LONDON

BARBARA J. KING is professor of anthropology at the College of William and Mary. She is the author or editor of many books, including *Being with Animals*. She writes regularly for National Public Radio's 13.7: *Cosmos & Culture* blog and reviews for the *Times Literary Supplement*.

The University of Chicago Press, Chicago 60637
The University of Chicago Press, Ltd., London
© 2013 by Barbara J. King
All rights reserved. Published 2013.
Printed in the United States of America

22 21 20 19 18 17 16 15 14 13 2 3 4 5

ISBN-13: 978-0-226-43694-4 (cloth)
ISBN-13: 978-0-226-04372-2 (e-book)

Library of Congress Cataloging-in-Publication Data

King, Barbara J., 1956–
How animals grieve / Barbara J. King.
pages ; cm
Includes index.
ISBN 978-0-226-43694-4 (cloth : alk. paper)
ISBN 978-0-226-04372-2 (e-book)
1. Grief in animals. I. Title.
QL785.27.K56 2013
591.5—dc23 2012045012

⊗ This paper meets the requirements of ANSI/NISO Z39.48-1992 (Permanence of Paper).

FOR CHARLIE, SARAH, AND BETTY
AND FOR CATS MICKEY AND HORUS,
GRAY & WHITE AND MICHAEL,
RABBITS CARAMEL AND OREO,
AND EVERY OTHER ANIMAL WE HAVE
LOVED AND LOST

EVERYTHING IS EVERYTHING
BUT YOU'RE MISSING.
BRUCE SPRINGSTEEN

CONTENTS

PROLOGUE ON GRIEF AND LOVE

One individual lies immobile, apart from the group. Everyone else rushes about, doing her work and keeping the high-functioning community running at top pitch. But the lone one lies dead—and ignored.

After about two days, a smell begins to waft from the body, a strong chemical odor. Soon, another individual comes by and carries the corpse to a nearby graveyard, where it joins many others—an efficient process of disposal. No one mourns.

Is this a scene from a zombie thriller, that often revived standby of Hollywood, Burbank, and, recently, the publishing industry? What real-life culture could treat its dead in this cold, mechanical way? Humans everywhere engage in elaborate rituals: preparing the body, comforting the bereaved, ushering the newly dead into an afterlife (or at least the cold, hard ground).

No, this graveyard scenario comes not from humans but from ants. Biologist E. O. Wilson observed the pattern in the 1950s: an ant dies, it lies ignored for some days, and then another ant comes and carries the body to the ant equivalent of a cemetery. The release of oleic acid from the ant's body, about two days after death, triggers the carrying response in other ants, Wilson told Robert Krulwich on National Public Radio in 2009.

Should a curious scientist borrow an ant, dab oleic acid onto its body, and return it to an ant trail, that ant—very much alive—will also be carried off to a graveyard, struggling all the while. Death-related behavior in these insects is, as far as we can tell, driven purely by chemicals.

While it's possible that entomologists just don't know how to recognize displays of insect emotion, I'm comfortable in hypothesizing that ants don't feel grief for their dead comrades.

Within the animal kingdom, ants are an extreme example. No one would expect a chimpanzee or an elephant to respond so mechanically to a whiff of chemicals. Chimpanzees and elephants are veritable "poster species" for animal cognition and emotion. Intelligent planners and problem-solvers, these big-brained mammals are emotionally attached to others in their communities. Finicky about with whom they spend their time, they may shriek or trumpet their joy when reuniting with preferred companions after a separation.

These animals do not just "exhibit social bonds," as the stilted language of animal-behavior science often suggests. The emotions that chimpanzees and elephants feel for others are closely bound up with their complex cognitive responses to the world. Chimpanzees are cultural beings who learn their tool-use patterns—fishing for termites, cracking hard nuts, or spearing bush babies in tree holes, depending on where they live—in ways specific to their group. And just like the old cliché, elephants never forget. They remember events vividly, to the point that they may suffer with post-traumatic stress disorder, as when their sleep is disrupted by nightmares after witnessing the killing of relatives or friends by ivory poachers.

Chimpanzees and elephants feel grief. Pioneering women field scientists Jane Goodall, observing chimpanzees in Tanzania, and Cynthia Moss, studying elephants in Kenya, reported years ago firsthand observations of the sorrow these animals felt at the death of loved ones. It's only natural, then, that chimpanzees and elephants appear in this book. The newest science adds fascinating new depth and details to Goodall's and Moss's original reports on grief in these species.

Animal grief is expressed and observed far beyond the African forests and savannahs, however. In this book, we will visit a variety of ecosystems to discover what is known about how wild birds, dolphins, whales, monkeys, buffalo, and bears—even turtles—mourn their losses. We will also peek into homes, and venture onto farms, in order to discover how our companion animals—cats, dogs, rabbits, goats, and horses—experience grief.

Historically, science has badly underestimated animal thinking and feeling. But now, scientists, often armed with videotaped evidence, are showing us that more animal species think and feel more deeply than we'd ever suspected.

Take goats and chickens, two animals whose potential for thinking and feeling I had, for years, barely given a second thought. How many times had I seen goats clustered in farms or yards, near my home in Virginia or on my travels in Africa, and yet not really seen them—and the same for chickens? Like most people, I create an implicit, mental hierarchy of animals when it comes to cognition and emotion. My working, if subconscious, assumption was that chimpanzees and elephants, on this scale, tower over animals like goats or chickens, who are just there in the background—or on our dinner plates.

Goat is the most widely consumed meat in the world and a dietary staple in Mexico, Greece, India, and Italy. It has also, over the last several years, been edging its way onto upscale plates in the United States. I have not eaten goat; I've been a near-vegetarian for a while now. Only recently, after hanging out with some nearby goats, corresponding with friends who have raised goats, and reading Brad Kessler's memoir *Goat Song*, have I begun to see goats as the complicated creatures they are.

I met Bea and Abby, mother and daughter goats of unknown breed, one sunny afternoon last year. They reside at the 4BarW Ranch, the home of Lynda and Rich Ulrich, near my home in Gloucester County, Virginia. When I met Lynda and Rich, I felt instantly that I was in the presence of like-minded souls. Rescued goats, horses, dogs, and a cat roam the ranch, and my hosts were full of the good stories that animal-rescue people love to exchange.

Bea is a pretty off-white shade, with a wispy beard and a calm manner; her daughter, Abby, is the same color but beardless. Lynda and Rich acquired Bea first, and only six weeks or so later did Abby join the other goats at the ranch, where they roam together through a large enclosure. When Bea and Abby reunited, they expressed what can only be called goat joy. They coo-vocalized, rubbed their faces together, and cuddled together in an explosion of mutual affection that brought tears to Lynda's eyes.

BEA AND ABBY. *PHOTO BY DAVID L. JUSTIS, MD.*

In his book, Kessler put it this way:

> The longer I spent with our goats, the more complex and wonderful their emotional life seemed: their moods, desires, sensitivity, intelligence, attachments to place and one another, and us. But also the way they communicated messages with their bodies, voices, and eyes in ways I can't try to translate: their goat song.

Greek tragedies were once known as "goat-songs," perhaps because goats were given to winners of Athenian drama competitions—and then sacrificed. When that happened, people offered a ritual song, but as we will see later on, goat voices too may lament a death.

Goats do not make tools like chimpanzees do, and it's probable that they don't recall past events or experience traumatic memories to the degree that elephants do. Their self-awareness is not as developed, and they wouldn't, for example, recognize their own images in a mirror. But should chimpanzees and elephants be the gold standard for animal thinking and feeling? Good animal-behavior science has forced us to rethink the tradition of judging apes' and elephants' ways of thinking and feeling by the nature of our own. It's no better a practice to judge all other animals by what chimpanzees and elephants do. Goat thinking and feeling is thinking and feeling.

Chickens, though? From childhood on into my fifties, I consumed hundreds of chickens. Poultry dishes were my favorites when dining out. The terms "chicken intelligence" and "chicken personality" struck me as oxymoronic, not reasonable description of chicken reality. All of this—the diet and the thinking both—have shifted as a result of stories from those who know better.

It started with Jeane Kraines, a friend who keeps chickens at her home in suburban New Jersey. She's had as many as fourteen at a time and has fallen into the habit of letting them roam the neighborhood to visit the neighbors. "Once I found them at a bridal shower," she told me, "all the lady guests in a circle around them. In the evening they would come back and I would close the door before nightfall."

The story I loved best from Jeane I call "the swimming pool rescue." One day, in her kitchen, she heard alarmed calls from the backyard, and the chickens rushed up onto her deck. "They were knocking furiously on the sliding door with their beaks," she remembers. "I ran outside immediately and they rushed off with me behind, trying to keep up. Straight to the pool we dashed. There I saw Cloudy, everyone's favorite hen, flailing her wings in the swimming pool. I reached in and lifted her out." Jeane is certain Cloudy's life was saved only by the resourceful action of her flock.

The sequence of steps taken by these chickens is remarkable. They recognized that a companion was in trouble; they knew where to seek help from the human world and how to get a human's attention; and in an embodied way, they directed that human immediately to the source of the trouble.

In its recounting of the smarts and social graces of "chickenkind," the book *Chicken* by Annie Potts rocked my world. Potts describes how chickens can do a lot of things we humans prize—they can recognize up to a hundred distinct faces, and a whole object when shown only a part. She is best when focusing on individual chickens, such as the charismatic Mr. Henry Joy who, by force of his personality, became a beloved therapy animal in nursing homes.

Potts touches on grief by way of a story borrowed from the zoologist Maurice Burton. One hen, old and nearly blind, was assisted by a second who was young and in fine shape. The younger hen collected food for her companion and helped the older one settle into a nest at night. Then the old hen died. The younger one stopped eating and weakened. Within two weeks, she too had died. Chickens think and feel. They grieve.

But the statement I've just made—that chickens grieve—is written in shorthand. It's more accurate to phrase it this way: Chickens, like chimpanzees, elephants, and goats, have a capacity for grief. Depending on their personalities and on the context, this capacity for grief may be expressed—just as is true for humans. It's possible to live with chickens or goats or cats and not witness any dramatic expression of grief when a member of the flock or herd or household dies.

Is it any different for humans? In its Metropolitan Diary column for January 16, 2012, the *New York Times* published an account by Wendy Thaxter about a day she and her sister were tending a community garden in Manhattan. A woman, known to neither sister, approached with a paper bag containing the ashes of her father. The woman asked if the ashes could be scattered in the garden, handed them the bag, and left, saying, "Here, please take this. His name was Abe, and I've had more than enough of him." We may laugh or gasp at her remark, but the point is, it's no use trying to predict how an individual will react to losing a relative or some other person who has played a role in her life. People may not grieve when someone close to them dies. Or they may grieve in an interior way, invisible to others, or only when alone.

In writing about animal bereavement, I walk a line stretched taut between two poles. The first is this wish to recognize the emotional lives of other animals. The other is my need to honor human uniqueness. I am, after all, an anthropologist. Anthropologists have described many ways in which our own species is unique in how we grieve. Just as chimpanzees aren't ants controlled by chemistry, we humans aren't elaborated chimpanzees. Among animals, we alone fully anticipate the inevitability of death. We grasp that, one day, our minds will fade, our breathing stop—whether gently or with horrifying suddenness, we can't know. We express in a thousand glorious or ragged ways our losses, losses of those we love.

When a child dies, a child who should have outlived us by decades, we may howl in sorrow, and some may labor to shape that howl into art. "Gut me," writes Roger Rosenblatt in a book about the sudden death of his daughter, a mother to three young children. "Slice me down the earth's meridian, from north to south. Lay my bones outside my skin." No other animal expresses grief like this, or attends death with ceremonies as varied as the languages of the globe. Since our ancestors first scattered red ocher on bodies many thousands of years ago, since grave goods were first offered to the dead for their use in an afterlife, since we invented tombs and cremations and sitting shiva, since we began commemorating death on Facebook and Twitter—across the millennia, we've come together to ritualize our mourning. We act in ways no other animal acts in the face of death.

Goat grief, then, is not chicken grief. And chicken grief is not chimpanzee grief or elephant grief or human grief. The differences matter. But differences between species may be rivaled by differences among individuals of the same species. The great lesson of twentieth-century animal-behavior research was that there is no one way to be chimpanzee or goat or chicken, just as there is no one way to be human.

We are alike, humans and other animals, and we are different. Balanced between these poles, I find the commonalities more compelling. I think this is because, like us, animals grieve when they have loved. We may even construe animal grief as a strong indicator of animal love.

Is it outlandish to write of animal love? How could we ever recognize what love is to an ape, much less to a goat? To describe fully what it means for people to love requires more than measuring the hormonal spikes in a besotted person's blood or charting the words, gestures, and glances shared by a new couple. Science can help measure love, but it can't tell the full story. Surely this challenge to science deepens when dealing with creatures who think and feel without language, or at best, with languages that lack words and sentences as we know them?

Noted animal behaviorist and animal-welfare activist Marc Bekoff acknowledges that the topic of animal love may provoke skepticism—which he counters in an exciting way. We have always, Bekoff observes, since we became people, grappled with the difficulty of defining or understanding love. "And yet," he writes, "though we don't truly understand love, we do not deny its existence, nor do we deny its power. We experience or witness love every day, in a hundred different forms; indeed, grief is but the price of love. Since animals grieve, surely they must feel love too."

Based on the science of animal emotion as explored by Bekoff, Goodall, Moss, and other scientists, I feel comfortable working from a platform of expectations about animal love that may also be seen as hypotheses to be tested in the future. Here's the central idea: When an animal feels love for another, she will go out of her way to be near to, and positively interact with, the loved one, for reasons that may include but also go beyond such survival-based purposes as foraging, predator defense, mating, and reproduction.

In the framework that I want to use, this active choosing by one animal to be with another is a necessary condition—a basic foundation for love. But it is only a necessary condition, not a sufficient one, to claim that we have identified animal love. Another ingredient is also needed:

Should the animals no longer be able to spend time together—the death of one partner being one possible reason—the animal who loves will suffer in some visible way. She may refuse to eat, lose weight, become ill, act out, grow listless, or exhibit body language that conveys sadness or depression.

For my definition to work, it has to distinguish, in most cases, between two types of situation. Consider a pair of wild chimpanzees, whom we will call Moja and Mbili, who travel together, rest together, and groom each other. They might do so because they feel some sort of robust positive emotion for each other and in each other's company. Or there might not be much attendant emotion. Maybe Moja and Mbili just fell into the habit of associating with each other but would be equally content with another female companion if the need arose. How could scientists figure out which of these two interpretations—if either—is correct? (In either case, the alliance may be beneficial in terms of acquiring resources; remember, survival needs are not excluded by this definition of love, but they must be supplemented by something more.)

Through careful observation and, optimally, analysis of film that documents the pair's interactions, we might be able to recognize love in, say, the intensity with which they seek each other out and embrace when they come together, or the care with which they groom one another.

But it would be a great mistake to too liberally apply the term "love" to animal relationships. Anthropomorphic excess may cause us to miss critical distinctions. And here is where our second, sufficient condition comes in. If Moja and Mbili feel love, one will show signs of grief when forced apart from the other, especially if the other dies.

Now, this two-part approach for evaluating grief in the animal world is not perfect. It may underestimate the extent of animal love, because the sufficient condition—separation or death—won't always be observable. Conversely, it's possible that an animal who doesn't love a companion may still feel grief when that companion dies. Another problem is that we may be unable to distinguish different varieties of love. If Moja and Mbili are mother and daughter, would their love differ from the strong feelings that might be shared, for example, by two females who met after migrating into the same community from different natal groups?

Of course, these distinctions may be hard to make even in humans. The love felt for one's family, one's friends, and one's mate or life partner may differ, and so may the grief that emerges in response to the loss of these various loves. But are these emotional differences visible to those of us looking in from the outside (as we must do with animals)? Only sometimes.

The arena of animal emotion offers significant challenges for observers. My definitional scheme is one place to start. Above all, we must always keep in mind the possibility that some animal love or grief will look quite different from our own love or grief, or the emotion of other group-living primates like chimpanzees, whose actions may resonate with us more readily.

As you read the stories in this book and watch the video clips that I suggest in the "Readings and Visual Resources" section, keep in mind the "ideal definition" of grief that I have offered, and how it relates to love. Some of the animals profiled in these pages do cleanly meet the stated conditions for love and grief, but not all. In some cases, we glimpse tantalizing hints of grief and love; in other cases, the relevant reports are too opaque to allow any certainty about what the animals feel. At this point in humans' quest to understand animal grief, however, even hints and opaque observations matter, because they will lead us to ask more insightful questions as we observe animals in the future.

After all the cautions and qualifications that I feel obliged as a scientist to offer, here's my bottom line: When we find animal grief, we are likely to find animal love, and vice versa. It's as if the two share emotional borders. Think of it as looking at one of those famous optical illusions. You stare at the drawing and at first it's clearly a bunny, but as you continue to look there's a visual shift, and suddenly it's a duck.

You'll find animal grief in these pages, in bunnies and ducks and a host of other species. But you'll also find animal love.

1 KEENING FOR CARSON THE CAT

The home of my friends Karen and Ron Flowe in Gloucester, Virginia, is decorated merrily. Candles shine a welcome in each window. An all-white Christmas tree graces the entrance, and a multicolored tree sparkles upstairs. The special music, food, and holiday anticipations of December have long enchanted the household.

This year, though, the air is scented with sorrow. Willa, a Siamese cat, wanders from room to decorated room, pausing first at the ottoman in front of the fireplace. With a glance at the soft, warm cushion, she lets out a wail. Moving on to the master bedroom, she jumps to the head of the bed and pushes her face and body into a cozy cave-space behind the pillows. She looks, and looks; another wail escapes her. It's sudden and terrible, not a noise one would expect from a cat.

Willa cannot still herself. The only thing that helps is when she's folded into an embrace by Karen or Ron, or when she's reclined in one of their laps. She is searching for her sister Carson, who died earlier in the month. For the first time in fourteen years, Willa is no longer a sibling, no longer the more outgoing and dominant half of an enduring partnership.

She's Willa, alone. And she grieves.

Willa and Carson—named for famed writers Willa Cather and Carson McCullers—arrived in this literary Virginia household on Shakespeare's birthday, April 23. Willa had been the plump pick of the litter. Carson had been offered at half price. Runty, a bit, the seller acknowledged.

WILLA AND CARSON. *PHOTO BY KAREN S. FLOWE.*

Carson acted in some peculiar ways right from the first week. Showing unusual sensory acuity, she fluffed up at the smallest sounds or movements. During one windstorm, she climbed to Ron's shoulder and pressed tightly against his neck. Sometimes, and strangely, she would progress across a room not in a linear path but by walking in circles. She did not meow, and even her purrs were faint; the Flowes concluded she was mute.

Then came the declawing. Willa and Carson went together for the surgery (like many of us, the Flowes routinely declawed their indoor cats for years but no longer subscribe to the practice). The veterinarian, however, missed one of Willa's claws, so back she went for follow-up care. Left at home, Carson began to scream, recalls Karen. Carson searched the house for Willa, and cried.

The sisters were soon reunited, and a contented life ensued, ruled by sun patches, excellent food, and loving laps. Always the leader, Willa would strike out for a favored spot—the warm ottoman or a preferred corner of the bed—and Carson would follow. Once settled, the pair often pressed their bodies lightly together, like the joined wings of a butterfly. If one fell ill, the other would tend her sister and groom her.

As she aged, Carson struggled with severe arthritis and developed a troublesome bowel-impaction problem. She lost weight and required surgery. Trips to the vet became routine. When Carson was away from the house, Willa acted out of sorts, but these separations were brief, and Carson always recovered enough from her bad patches to sustain an active relationship with her sister.

Then, on a December day, Carson began to shake, a symptom she had not experienced before. Her body temperature dropped, and at the vet's recommendation, she was put into an incubator. Cocooned in the incubator's warmth, Carson went to sleep that night and never woke up.

Ron and Karen felt grateful that Carson had died in her sleep, with no suffering. Yet their grief was substantial. As for Willa, she at first expressed that mild, something's-not-quite-right, out-of-sorts mood that one sister typically exhibited when apart from the other. The Flowes anticipated a stronger response might ensue, but they were unprepared for what did happen.

"Within two or three days," Karen says, "Willa started acting bizarrely. She was looking and looking for Carson, and she began emitting these sounds that we had never heard from her, that I have never heard from any animal, as a matter of fact. If I'm going to be literary, I've read Irish literature and it talks about the keening for the dead—and keening sounds like the closest to what she was doing. Willa was searching all the time, and all of a sudden she'd let out this horrific . . ." Karen's voice trails off, then picks back up: "As soon as Willa got into my lap, she would stop. She was grieving. She is getting better now, just like humans do."

Was Willa expressing grief? Couldn't it be that she was just unsettled by the sudden change in her day-to-day life? Writing for the magazine *Modern Dog*, Stanley Coren remarks on just this point, one that applies equally to cats: "In the animal behavior world at large, the jury is still out on whether dogs are actually mourning the loss of a loved one, or simply exhibiting anxieties related to the change in routine."

Skeptics love to cry "Anthropomorphism!," suggesting that animal lovers too readily ascribe human emotions to other creatures. And the skeptics have a point: Rather than accepting uncritically the existence of animal grief, or animal love, or any other complex emotion in non-human animals, we should first weigh other, simpler explanations. In Willa and Carson's case, we know something relevant from their long history together. We know that after the declawing surgery, Carson cried out for Willa even though Willa was out of the house only temporarily.

But Willa's response after Carson's death was an order of magnitude different from anything that had happened before. Karen is convinced that Willa had intuited a finality in her sister's absence. In part this may have been triggered by Karen and Ron's own mourning, visible and audible as it was to Willa. And in part, too, it may relate to the choices that the sisters made, day after day, to intertwine their bodies. Mightn't that physicality have led to an embodied sort of knowledge? Wouldn't Willa somehow grasp, when she no longer could curl up with her sister, that Carson's absence was permanent?

I want to emphasize that animal grief does not depend upon a cognitive mastering of the concept of death. That's a recurring message of the stories and the science in this book. We humans anticipate—sometimes dread, sometimes welcome—death, and after a certain point in childhood, we grasp what it means to die. Maybe some other animals do have a sense of that finality, as Karen posits for Willa. Yet as I noted in the prologue, my definition of grief is linked not to some feat of thinking but instead to feeling. Grief blooms because two animals bond, they care, maybe they even love—because of a heart's certainty that another's presence is as necessary as air.

When it came to Carson, Willa's heart bore that certainty. Karen wondered what to do for Willa the survivor, other than dosing her with extra affection. She mused about acquiring another adult cat, in order to restore the household's Siamese symmetry. But she knew a powerful if simple truth that cuts across species lines: loved ones are irreplaceable.

In *Histoires Naturelles*, a nineteenth-century collection of essays devoted to animals of the French countryside, Jules Renard writes of an ox named Castor. One morning, Castor emerges from his shed and heads as usual to his yoke. "Like a maid drowsing, broom in hand, he goes on chewing while waiting for Pollux," his longtime partner.

But something has happened. What precisely it is, Renard does not say. The dog yaps nervously. The farmhands run about and shout. And next to him, Castor feels "twisting, hitting . . . fuming." Turning to look, he sees not Pollux but another ox. "Castor misses his partner," writes Renard, "and seeing the troubled eye of this unknown ox next to him, he stops chewing."

How much feeling Renard crams into this understated passage. Castor isn't content with just any ox; it's Pollux he knows, Pollux he misses. Animals matter to each other as individuals. Sisters matter.

Eventually, Karen and Ron did adopt a young cat named Amy, a gorgeous Russian blue with a pretty "locket" of white hairs on her upper chest. Amy had been left at an animal shelter by a breeder of pure Russian blue cats; those few white hairs, not standard for her breed, reduced her value. (Amy's rejection by the breeder strengthens my resolve to adopt animals from shelters or other rescue organizations, as we have done with our six indoor cats.) When Karen visited the shelter in search of a companion for Willa, Amy climbed into her lap and settled into a purr. And just as Amy chose Karen that day, Karen chose Amy.

In bringing Amy home to Willa, Karen hoped to tap into a phenomenon that animal-behavior scientists discovered decades ago: When emotionally troubled, a social animal may reap great benefits from caring for a younger companion. This principle was driven home in the aftermath of the "separation experiments" done in the 1960s by Harry Harlow and his colleagues, who sought to understand the nature of the mother-infant bond in monkeys and what happens in its absence.

These scientists famously demonstrated that six months' or a year's worth of isolation caused young rhesus monkeys to become psychologically disturbed. Without the company and comfort of their mothers or other companions, the monkeys rocked back and forth, clasped themselves, and acted exactly like what they were, severely depressed primates. It's painful to read about these experiments now, because the monkeys suffered so much to prove a point that in retrospect looks startlingly obvious.

When introduced to other monkeys their age who had been reared normally, these disturbed monkeys couldn't cope well. Lacking social experience, they knew nothing of the right signals to give their peers to bring about positive encounters. But when given a chance to spend time with normal younger monkeys, even the harmed, broken, motherless monkeys began to improve. The younger monkeys functioned, the scientists found, as therapists of a sort. In a classic paper published in 1971, Harlow and Stephen Suomi wrote, "6-month-old social isolates exposed to 3-month-old normal monkeys achieved essentially complete social recovery."

What the monkey experiments demonstrated is the balm of responding, even when in emotional pain, to creatures who are younger and less threatening. Of course, Willa was not a social isolate, so the analogy with the monkey experiments only goes so far. But the idea is broadly similar. When Amy first arrived in the house, Willa vocalized her objection. She made a noise wholly unlike her Carson-wail, a growl more like the roar of a tiny lion. And with it, the point was made: Willa was not thrilled to have this unfamiliar cat, however young and unthreatening, on her turf.

Soon, though, Willa began to engage more actively with what was happening around her than she had in months. She actively sought to be in the same room as the new arrival. "Willa had something new to think about," Karen told me with a smile. Even though her first response wasn't a warm one, Willa was coaxed by Amy's presence out of her diminished state, the condition of mild emotional detachment in which she had dwelled since the loss of Carson.

At first, Willa and Amy, even if in the same room, kept their distance. Only in one situation did they tolerate each other's nearness: when both wanted to be physically close to Karen. When Karen relaxed on the couch or reclined in bed, the two cats took up positions on either side of her—safely separated by the person they loved. This went on for about six months. Then, one fall day, Karen fell asleep on the couch with Willa snuggled close. After perhaps an hour's nap, Karen awoke to find Willa still in place near her hip, and Amy up at her shoulder. The two cats were in fur-to-fur contact. "And," as Karen puts it, "there were no ugly noises!"

Willa and Amy's relationship entered a fresh phase. Once, Amy licked Willa from head to toe; Willa expressed no blissful purr, but she did allow the intimacy. And then the two cats began eating side by side, from the same bowl. The relationship that Willa and Amy have developed is far less intimate than what Willa shared with her sister for all those years. Willa and Amy never knot themselves into a tight circle or press close in the siblings' butterfly formation. Willa has selected a new favorite sleeping place, one she never frequented when Carson was alive. She burrows into the space between Karen's pillow and Ron's, and faces the wood of the headboard. Once, Karen caught Amy investigating this spot, as if to see what the attraction was for Willa, but Amy never tries to sleep there.

A remnant of that tight two-sister circle remains. Willa, when napping on the bed or on the ottoman (a spot she shared with her sister), still makes her half-moon. To Karen, this is an evocative image because it's so incomplete: the empty space speaks to her of Carson. "There's an emptiness in Willa's posture now," Karen says.

Karen knows that Willa's physical and emotional well-being have improved since Amy's arrival. Willa has put on weight, grooms herself more fastidiously, and is more vigorous in her overall approach to life. Are memories of Carson still rooted in Willa's mind? Do images of sharing the fire-warmed ottoman with her sister flicker through her dreams? This realm of the cat-mind goes beyond science.

In 2011 I began writing a weekly post on anthropology and animal behavior at National Public Radio's 13.7 blog, which is devoted to science and culture. In a piece about animal grief, I offered a short version of Willa's and Carson's story. Back came responses about readers' own experience with animals who mourn.

Kate B.'s story has powerful parallels to Willa and Carson's. For fifteen years, two Siamese cats, brothers named Niles and Maxwell, lived with Kate's parents. Niles became ill from pancreatic cancer, and when it came time to put him down, Maxwell accompanied his brother to the vet. Soon, Maxwell found himself back home, surrounded by familiar places and favorite things, but without his sibling.

"Maxwell spent the next several months," Kate remembers, "constantly wandering the house and crying the most agonizing cries, looking for his brother." As it turned out, Maxwell lived just several more months. During this time, he derived the most comfort from the visits of Kate's three young cats, who were brought to see him and who bonded with him. Now, when Kate brings the trio to her parent's house, they search for Maxwell, and one of them often sleeps exactly where Maxwell slept.

Sibling ties, like those between sisters Willa and Carson and brothers Niles and Maxwell, are powerful connections that, when broken, may give way to mourning. Cats may mourn a lost companion, though, even when no blood tie exists. Channah Pastorius described the friendship between Boris, a brown tabby she adopted from a shelter, and Fritz, a kitten her son brought home. The two felines play-wrestled and slept with their front legs entwined. At age eight, Boris developed kidney failure, but with good veterinary care and lots of TLC, he managed to live two and a half more years.

Finally the inevitable came to pass, and Boris was put down. What happened next has a familiar ring. "Fritz mourned the loss of Boris with sad yowls and depression," Channah wrote. Fritz was lethargic, uninterested in his favorite toys or much of anything else.

But a second common thread links Fritz's story with those of Willa and Maxwell: when a tiny black kitten showed up on the family's patio and ran indoors, Fritz perked up. He began right away to play with the kitten, named Scooter by the family. Once again, a new and younger partner abated at least some of the grief.

Mourning may cross even species lines, as we will see again in later chapters. Kathleen Kenna's fifteen-year-old cat Wompa reacted strongly when the family dog, eight-year-old Kuma, died after a long illness. Wompa had regularly allowed the dog to groom her, and the two acted like best buddies. (Meanwhile, the other cat in the house had nothing to do with the dog.) A few days after Kuma succumbed to cancer, Wompa began to moan loudly. The strange, intermittent crying, which sounded "like a banshee" to Kathleen, lasted for several days. The cat also shifted her nighttime sleeping place to the spot at the end of the bed where Kuma had slept.

After I discussed animals' responses to death on a radio program, listener Laura Nix e-mailed me about two cats called Dusty and Rusty, who had lived with her friends for many years. They were sisters, but they were no Willa and Carson! Their relationship was downright antagonistic, to the degree that they carved up the house into two territories:

Dusty lived upstairs, and Rusty lived downstairs. When Dusty began to fail, in her old age, she was cared for lovingly by Laura's friend. On the night that she died, indeed, at the moment she died, Rusty—who was, as always, downstairs and apart from her sister—let out a single howl. Laura notes, "It was the only time I ever heard her make such a sound. I can't tell you how she apparently knew."

Although I live surrounded by cats and am attuned to the possibility of animal grief, I have never witnessed cat mourning. We have lost cats to illness and old age, but the only emotional disruption in the household came from our own grief. Perhaps part of the explanation is that the cats we lost were primarily attached to us rather than to the other cats. Our cats are rescues, and we have a lot of them. Six live indoors with us, and twice that number reside in a spacious pen in our yard. Nestled under trees, sturdily built, with a two-story cat hotel and other hidden grottoes for warmth and shelter, the pen offers sanctuary to these cats, most of whom had lived as part of a feral colony at a public boat landing on the York River, not far from our house. At one point, a few people, annoyed by the colony's presence, threatened to harm the cats. Building the pen was my husband's answer to that threat. As hard as we work to reduce the feral-cat population to zero through spay-neuter programs, we want to help the cats who need us right now.

We enjoy the company of these small creatures, no longer have to fend for themselves against hunger, dogs, coyotes, and uncaring humans. When I walk outside and enter the pen, I enjoy watching shy Big Orange sleep soundly under a bush, one-eyed Scout jump at a bug, and friends Dexter and Daniel relax together near the picnic table. Haley and Kaley, nicknamed "the white sisters," have never been feral; when a friend called urgently seeking someone who would adopt the two together, before they were euthanized as unwanted, we took them in. These siblings are the most closely bonded of any cats in our care. Kaley is a bit heavier than her sister, with one eye blue, the other green. Haley has a darker smudge atop her head and "talks" more to us humans. The sisters seem hyperaware of each other's location in the pen and most often choose to eat, rest, or bask in proximity to each other. We don't know their precise ages, but they've been together since birth, at least three or four years. Haley and Kaley are far closer to each other than any other pair of cats we have had. What will happen when one of the white sisters dies? I hope we won't find out for many years.

Clearly, I'm cat-preoccupied. Yet there's another reason why I chose to launch this book with a chapter on cat mourning. Words like "aloof" and "independent" are often used to describe cats' personalities. When my former dean at the College of William and Mary joked that trying to achieve consensus among faculty members is "like herding cats," everyone laughed. Immediately, we grasped the cross-species analogy. It's an old stereotype, pitting independent, almost rogue cats against ultraloyal, comparatively tractable dogs. And there's some truth to it: dogs evolved from pack animals and are, on average, more attuned to humans than cats. But individual cats, depending on their personalities, may bond with other cats and with people just as deeply as dogs bond with other dogs and with people. And when death comes for one cat, that bonding may lead to mourning for the survivor.

Willa is a survivor. By all appearances, she enjoys Amy's company. Even so, Amy isn't Carson. Willa lives on without her sister—but in a very real sense, it's a sister that Willa remains.

2 A DOG'S BEST FRIEND

Grief is often born from love. That's the single most beautiful thing I've come to appreciate through more than a year's immersion in reading and writing about bereavement.

With dogs, love is often easy to see, especially the whole-body love they share with us. Energy courses through every muscle to wag the body as much as the tail; liquid eyes overflow with their joy at our company. Dog love is entwined with loyalty, a canine trait that's the stuff of legends.

Visitors to Tokyo sometimes make a pilgrimage to the Shibuya train station to view a statue of an Akita dog named Hachiko. Hachiko, or Hachi, as he was nicknamed, was born in 1923 and adopted soon after by Eisaburo Enyo, a professor at Tokyo's Imperial University. Every day the professor walked from his home to Shibuya station to board the train that took him to his office. And every day, Hachi trotted by his side. When Mr. Enyo's morning train departed, the dog returned home—only to return later, to meet the evening train.

For over a year, this was their pattern. Then Mr. Enyo died, suddenly, at his university office. Hachi was left waiting at the station for his friend who would never again come home. For more than ten years, he continued this ritual, going each morning to Shibuya station and waiting quietly. Even in old age, when he moved more stiffly, Hachi maintained his vigil, seeking the one face that mattered to him. In 1935 Hachi died. Every April 8, dog lovers honor his memory during a ceremony at the train station, held in front of Hachi's statue. A Japanese film based on the story was released in 1987; a 2009 American remake stars Richard Gere as Mr. Enyo.

I tear up when I think of Hachi: nothing dissuaded this dog from his loyal and hopeful waiting. He remembered his friend and acted, not as if grieving or depressed, but in a wholly purposeful manner, as if he expected to see him again at any moment. Most of us, I think, long to matter to someone as much as Mr. Enyo mattered to Hachi. Just as importantly, we long to be remembered after we are gone as Hachi remembered Mr. Enyo. In Alexander McCall Smith's novel *The Charming Quirks of Others*, the character Isabel, a philosopher, quotes a line from Horace to her partner Jamie: *Non omnis moriar*, I shall not wholly die. She then remarks, "Only if there were nobody at all left to remember would death be complete."

Hachi's story is about love and loyalty across species lines. But what about dogs' interactions with each other? We regularly see the dogs we live with or encounter around town playing joyfully together, or slipping back and forth between mild conflict and relaxed companionship. But is there genuine love and loyalty among dogs?

Time magazine ignited a small firestorm among dog lovers with a cover that blared "Animal Friendships" above a photograph of a big brown hound and a tiny white Chihuahua. The story itself was dismissive of the notion that dogs share enduring friendships. Dogs' interactions lack the "constancy, reciprocity and mutual defense observed in species such as chimpanzees and dolphins," wrote journalist Carl Zimmer. At this, dog people got hot under the collar. Led by animal behaviorist and dog trainer Patricia McConnell, they fired back that scientists underestimate dogs.

Acts of dog-to-dog loyalty certainly exist. A video recorded in Chile shows cars and trucks whizzing by on a multilane highway. In the middle of the road a dog lies motionless. It has apparently been hit by a vehicle and is severely injured, if still alive. Then into the frame comes a second dog, zigzagging through the traffic. This dog isn't very big, but he (or she) exudes a sense of determined purpose. He safely reaches the injured dog and begins to drag him to the median, even as cars zoom past. As a rescue worker approaches the two dogs, the video abruptly stops.

The video's narration is in Spanish, but none of us can fail to see what has happened: One dog has risked his life for another. From ensuing newspaper reports, we know that the injured dog died and the second dog ran off. Here again, as with Hachi and Mr. Enyo, we see no expression of dog grief. But in this case we have no back story, no idea of the circumstances. Were the two dogs friends with a shared history? Were they related to each other? No one knows. The rescuer dog was never found, a disappointing outcome for the people who expressed a desire to adopt him. Nonetheless, as one newspaper put it, that brave dog elicited "worldwide admiration" once the video clip went viral. I like to think that this incident in Chile invited millions of people to reflect on the depth of dog emotion.

A happier outcome for a dog hit by a car is reported by Stanley Coren in *Modern Dog* (a publication with the delightful tag line "the lifestyle magazine for modern dogs and their companions"). Mickey, a Labrador retriever, and Piercy, a Chihuahua, were fast friends who lived together in a family's home. Mickey was the older, and of course, much bigger of the pair. One day, Piercy ran into traffic and was struck by a car. The family, crying over the body, placed their pet in a sack and buried him in a shallow grave in their garden. Even Mickey seemed to express sorrow; the big Lab sat at the grave even after the rest of the family retired to bed.

Sometime later, the father of the family was awakened by unusual noises coming from outside. He thought he heard a dog's whining. Going outside to investigate, he found an open grave and an empty sack—and Mickey feverishly attending to his small friend. As the man watched, Mickey licked Piercy's face and nuzzled Piercy's body. He carried out these actions with great energy, as if trying to revive Piercy. The man thought these attempts hopeless. But in a flash his certainty faded: he saw a spasm run through Piercy's body, then watched, astonished, as Piercy lifted his head and whimpered.

Mickey's acute hearing may have picked up sounds coming from Piercy's grave as the little dog found himself buried alive—cues that the man, or any other human being, could not possibly have heard. Or maybe it was the legendary canine sense of smell that tipped Mickey off. Whatever sensory capacities may have been involved, Mickey's love and loyalty must be fitted into the explanatory mix. If the big Lab hadn't felt such a bond with his tiny friend, he wouldn't have stayed vigilant by the grave, nor would he have labored so hard to extricate and revive the Chihuahua. Without Mickey, Piercy would surely have suffocated.

Reports of heroic acts like those carried out by Mickey and the dog in Chile come along rarely. But they point toward a capacity for the emotions that may underlie dog grief. When two dogs enjoy each other's company and are keenly attuned to each other's whereabouts, actions, and moods, the conditions are right for one to mourn when the other dies.

Almost no scientific research has been carried out on dog grief. A recent wave of studies into aspects of dog cognition, however, supports the notion that dogs are incredibly sensitive to others around them. In a series of experiments, psychologists Brian Hare and Michael Tomasello found that domestic dogs outperform chimpanzees at comprehension of human gestures. The test they describe is beautiful in its simplicity: A person hides a desired food or object in one of several opaque containers, then points or directs her gaze toward the baited container. The question is, will a watching animal follow this cue and head straight for the correct container in order to grab its reward?

If the animal in question is a human, and over fourteen months of age, the answer is consistently yes. The same is true for domestic dogs, who make a beeline to the container concealing the treat. In fact, dogs succeed even when the experimenter complicates the task by standing a meter away from the containers and pointing with the cross-lateral hand, or pointing to the right container while walking toward the wrong one.

Chimpanzees don't do nearly as well on these tests. The key to success for at least some nonhuman animals seems not to be pure brain power, but instead a lengthy period of mutual attunement with humans. Thanks to their history of domestication, dogs have had extensive "practice" reading the movements of human companions. DNA science, together with archaeological research, tells us that dogs and humans initiated this process over ten thousand years ago, maybe even as early as fifteen thousand years ago. The first domesticated dogs probably came from China or the Middle East, but the human-dog bond in prehistory was also widespread across Europe and Africa. When the first settlers crossed the Bering Strait into North America, they had dogs as their companions.

This exquisitely attuned dog-human relationship, set in motion by the domestication process, also affects what happens between dogs themselves. Of course, dogs evolved from wolves, animals with strong, pack-oriented social tendencies. The combination of biology and socialization has a powerful effect. In this regard, Hare and Tomasello report an exciting result: On the hidden-object test, dogs do equally well whether cues are provided by humans or by other dogs. Though I'm not certain how dogs indicate to other dogs the baited container, the take-home point is clear: Dogs are incredibly attentive to other dogs. What sometimes gets overlooked in research like Hare and Tomasello's is the emotion that may be woven through such acute dog-to-dog attention.

From the dozens of stories that have come my way about dogs mourning other dogs, a powerful trio of characteristics can be distilled: love, loyalty, and smarts. Often, the tales are shared by a person who loved a dog who has died and now worries about the emotional health of a surviving dog in the household. Such was the case with a member of my own family.

For sixteen years, Connie Hoskinson lived with a tiny silky terrier named Sydney. Connie and Sydney strolled through their suburban Virginia neighborhood for forty-five minutes every day, greeting friends and neighbors as they went. Back at the house, though, there was no question but that Sydney preferred the company of Connie's husband George.

When George's health began to fail, Sydney's devotion only increased. Near the end of his life, George could no longer rise easily from the sofa or bed, and Sydney altered his activities to stay with him. He brought his toys directly to George's lap, timed his naps to George's so the two would curl up together, and followed George to the bathroom, returning to lie down only when George did.

Then George died. The change was a tough one for Sydney, as it was for Connie. "For almost a year," Connie told me, "Sydney didn't have much to do with me. After that, he became my dog." Sydney's vibrant personality brought much pleasure to Connie. Even in the early days

SYDNEY AND ANGEL. *PHOTO BY CONSTANCE B. HOSKINSON.*

with Connie and George, he had enjoyed sitting on the piano stool and playing notes. When a person would play, he'd "sing" along. Later, after George was gone, he became closely attuned to Connie's moods. When Connie cried, her habit was to put her hands up to her face. In a show of concern, Sydney would try to pull her hands down.

When Sydney was thirteen years old, Connie adopted a second dog, an adult Maltese named Angel. Sydney's life changed again. He took great pleasure in the company of his new friend—so much, that he defected from his nightly sleeping spot on Connie's bed and joined Angel in the kitchen. For three years the pair slept side by side, Sydney in his blue dog bed and Angel in her pink one.

Then, suddenly, Angel died of a heart attack. Terribly upset, Connie awaited the arrival of a neighbor who would help her bury Angel. She placed Angel's body back in her small pink bed. Sydney crawled into that bed and lay with his head on Angel's still form.

Angel was buried that day. For the next three weeks, Sydney searched the house for her. One time, Connie found him in the laundry room, where Angel's bed was waiting to be washed. Sydney had pushed the bed over, apparently looking for Angel. Soon, he began to eat poorly, a pattern that only worsened over the three weeks. Despite counsel from a vet, and Connie's constant affection and offers of all kinds of food, Sydney lost weight. The thinner he became, the more worried Connie became.

Sydney had resumed sleeping in Connie's bed, reverting to the pattern of his pre-Angel days. One morning, Connie awoke to find that he had died during the night. She believes that Sydney just couldn't survive the loss of Angel.

Connie recalls what happened when she next took a neighborhood walk—her first one alone in sixteen years. "When my neighbor walked out to the street to meet me," she told me, "he was holding out his arms, because he knew for me to be walking alone, it had to be that my constant companion was gone."

To witness the mourning of a surviving animal can be a second blow, following close on the death of a pet. An online discussion of dog grief brought forth a cry for help in this regard. A woman's eighteen-year-old dachshund, Ginger, had been euthanized on a vet's recommendation. After fourteen years shared with Ginger, the woman felt her loss deeply. Her sad state was compounded when her second dog, an eight-year-old who had lived with Ginger since the age of six weeks, began to decline, much as Connie's Sydney had. It got to the point where the younger dog, Heidi, simply refused to eat. Her sleep was terribly disrupted as well. Ginger and Heidi had always eaten together, the woman reported, and they loved the same treats. "The treats are the only thing now that Heidi has even a slight interest in." The woman sought help: how could she help ease Heidi's grief?

In response to queries of this kind, *Modern Dog* magazine offered tips for dog owners in a sidebar to the article in which Mickey and Piercy were featured. Statistics from the ASPCA's Companion Animal Mourning project indicate that two-thirds of dogs exhibit negative behavioral changes after losing another dog from their household; these changes may linger for up to six months.

Loss of appetite, lethargy, and anxiety behaviors such as pacing and being "clingy" are the main changes for which a dog's caretaker should watch. Offering a grieving dog a regular exercise regime, enrichment such as toys and treats, and renewed training to provide extra routine and structure, all may help. Drugs such as Elavil and Prozac, the magazine suggests, may be needed in severe cases.

In tandem with exploring the sensitive emotional nature of dogs, we might ask some questions that depart from the scientific mainstream. Can dogs not only feel the death of a loved companion but somehow intuit when it is about to occur? Queries like this one are common among dog people, ranging from the sober-minded, "show me the evidence" types to those who embrace more inferential, "New Age" modes of thinking.

A few years ago, a caller to a radio program I was on recounted how one night, her dog, a dachshund, became agitated, vocalizing and behaving in unusual ways. The next morning, the caller learned by telephone that one of the dog's puppies—now living with another family—had died the previous night.

Could the dachshund's atypical behaviors be explained by some intuition of her pup's death? How, short of telepathy (an ability about which I am acutely skeptical), could a dog possibly come by this knowledge? The mother was separated not only from her pup but also from any person who had knowledge of the death. Here we venture deep into contested territory. It's a surprisingly popular claim that some dogs just "know things," as when they predict with great precision (by their excited behavior) their owners' return from work or travel. This precision holds, the claim goes, even when the return is unexpected or happens at an irregular time of day.

Events videotaped by the researcher Rupert Sheldrake show that dogs really do behave with excited anticipation when their owners start toward home from a distant location. One camera recorded the movements of a British woman called Pat Smart, while a second recorded those of her dog, Jaytee. Even when Sheldrake's researchers varied the timing of Smart's movements and controlled for other factors that might have caused Jaytee to become excited, the dog, through his behavior, indicated an awareness of Smart's setting off toward home. His alertness level shifted quickly and he began to look for Smart out the window. (For detailed analysis, see chapter 9 of my book *Being with Animals*).

Even when confronted with evidence of this sort, I'm wary. How could I not be? Scientists aren't much in the habit of accepting tales that rest on concepts uncomfortably close to animal ESP. Rigidly controlled research on more dogs is badly needed. And not only on dogs, either.

Oscar is a cat who is said to predict when elderly people in a Rhode Island nursing home are about to die. When Oscar curls up on the bed of a sick resident, staff members telephone the family to say that their loved one's death appears to be imminent—because the cat is just that reliable. David Dosa, the doctor who first described Oscar's behavior in the *New England Journal of Medicine*, later wrote a book about this unusual phenomenon. Sometimes, given the age and poor health of the nursing home's population, Oscar was forced to split his attention. If two residents neared death at about the same time, Oscar would stay with one till the end, then race on to the next. He wasn't prone to lingering with the body. Though his presence did give comfort to the families of dying patients, his behavior centered not on expressing grief but on detecting the near approach of death.

As Oscar makes clear, acute sensitivities in our pets are not confined to dogs. The explanation for Oscar's death predictions lies, I believe, with the smell of molecules called ketones as they are released from a dying body. This medical explanation does not diminish the fact that Oscar is a remarkable animal; it may be that his nose is not unusual, but his unique way of responding to what he smells is.

Our detour away from dog grief underscores the idea that domestic animals pay keen attention to what is happening around them. Yet just as not every smart cat is Oscar, not every dog mourns when confronted with death. We shouldn't fall into the trap of making universality a criterion for the existence of a phenomenon—by which I mean, we shouldn't require every dog to grieve in order to believe that some dogs do.

An intriguing comment, made as part of another Web-based discussion of dog grief, illustrates this variability in behavior:

> When I had to have Number One Dog put to sleep, Number Two Dog went to the vet's office too and was given an opportunity to see her best bud's remains. She was decidedly not interested, and I felt like a bit of a fool for anthropomorphizing her. I have no idea whether she ever "got" death. In fact, I suspect she didn't have a clue. The body she saw at the vet's office wasn't her lifelong friend; it was a thing she didn't know.

Perhaps some dogs simply lack the brainpower to make the connection between a dead body and a living, loved friend. I'm suspicious of this interpretation, however. Dog Two Number may well have been aware that the death had occurred, and may even have recognized the body, but nonetheless been indifferent to the other dog's death. Indeed, the owner remarked, being the only dog was very much to Dog Number Two's liking; this new status meant that she received more attention, and thus her life improved, a far more important outcome, for her, than the first dog's death.

Whether or not Dog Number Two recognized her late companion, reliable eyewitness reports strongly hint that some animals, from elephants to chimpanzees to bison, do recognize that a body represents, in changed form, a fellow animal who was once very much alive. We'll come to these stories later in the book.

Finally, a powerful photograph and accompanying tale about the actions of one Labrador retriever compel us to think hard about what mental connection dogs may make when a death occurs. In summer 2011, thirty American soldiers serving in Afghanistan were killed when the Taliban shot down their Chinook helicopter with a rocket-propelled grenade. Amid this enormous tragedy, one dog caught the nation's attention.

Hawkeye was the dog of US Navy SEAL Jon Tumilson, age thirty-five, one of the soldiers to die in the helicopter crash. Hawkeye had been a constant presence in Tumilson's life for years. When it came time for Tumilson's funeral, held in a school gymnasium in Rockford, Iowa, and packed with fifteen hundred mourners, Hawkeye was included. In fact, he led the family down the aisle toward the flag-draped casket. When a close friend of Tumilson's stood to eulogize the soldier, Hawkeye did something no one expected. He followed the friend to the front of the gym, lay down in front of the casket, and stayed there for the duration of the service. A photograph captured the solemnity of the occasion and the dog's fixed presence at the coffin.

Skeptics might offer alternative hypotheses to account for the dog's choice of a position right in front of the casket: Maybe it was just a coincidence, a comfortable place to rest, and Hawkeye had no comprehension that his dearest friend occupied the casket.

I prefer to take a second detour, around such objections. I am thinking instead about Hawkeye in the context of everything that we know about dog love, loyalty, and cognition, stretching back eighty years to the actions of a dog in Japan. I am reflecting upon Hawkeye's love for Jon Tumilson. And following this route, I come to know something and know it to a certainty: Whether Hawkeye grasped that Tumilson was in the casket isn't the key to understanding Hawkeye's grief, or any dog's grief.

When loyal dogs grieve, for a person or for another dog, they grieve because they have loved.

3 MOURNING ON THE FARM

Storm Warning was a beautiful thoroughbred with a challenging personality. So many things spooked the horse: umbrellas, bicycles, small dogs, ponies, even people who removed an item of clothing while riding him. Storm, as he was called, was just a bit neurotic. But he lucked out in one way: he enjoyed a fifteen-year close relationship with Mary Stapleton, who happens to be a psychologist. Acutely attuned to people's fears and anxieties, Mary transferred her insights and calming abilities to the horse. Even as Mary and Storm competed in the dressage ring, they worked together on Storm's fears. In Mary's words, Storm "learned to jump and face all of his terrors with great courage."

Then, one night when he was eighteen years old, tragedy struck. Storm had been turned out into a field at the farm where he lived in a herd with other geldings. An accident of some sort must have occurred, for in the morning, Storm was found to be severely injured. Examination revealed a compound fracture in his hind leg, too extensive for successful treatment. Right there in the field where he had spent his happiest days, Storm was put down. And right there he was buried.

Horse people will recognize, Mary says, how unusual it is for a horse to be interred in the fields where he had lived. Mary still expresses gratitude to the farm's owner for affording Storm such a burial.

The evening after Storm's death, Mary walked out into the field alone. Approaching the large mound that now covered the horse's remains, she placed on the ground his favorite flowers—flowers he used to eat. "I heard the horses grazing around me," Mary says, "and was, as always, comforted by their presence. Slowly, at least six of the group stood around the mound, stopped grazing, and looked at the grave. I realized we, the horses and I, had formed a circle around the fallen Storm."

To Mary, this event felt eerie, all the more so once she realized exactly who the encircling horses were: Storm's companions, the geldings who were part of his herd. The geldings stood with lowered heads, which implied a straight-ahead gaze. "If horses hold their heads high," Mary explained, "they are scanning far away. But Storm's group clearly was at the right visual angle for looking directly at the burial site." Other horses, nearby in the field but new to the farm and not part of Storm's herd group, did not join the circle. None of the gathered horses ate the flowers Mary had placed on the grave, and she had brought no other treats. Whatever drew Storm's companions to his burial place, it wasn't the hope of food. In spontaneously forming a circle at his grave, Storm's herdmates began a vigil of sorts; Mary found them still there the next morning. A cautious person, Mary acknowledges that many interpretations of this behavior are possible. "I choose to think," she says, "that I was allowed to share a circle of mourning for our mutual, loved companion."

Mary's story about Storm became a catalyst for me, as I was unfamiliar with horses either personally or in my work on animal emotion. To horse people, I soon found out, the notion of a horse circle, or indeed of equine grief, was anything but new.

At one time, Janelle Helling managed a ranch in the Colorado mountains, with twenty or thirty horses in residence. One morning, the herd failed to make its way to the barn-corral area for feeding as it usually did. A mare had foaled during the night, and the newborn was too weak to stand. "The rest of the horses were circled around the mare and foal," Janelle recalls, "and would not let us get near them. The horses refused to be herded away from acting as a barrier between us and the mare and foal."

That barrier was protective in nature. In that area of Colorado, mountain lion, bear, and coyote are indigenous, so perhaps the horses were hypervigilant for predators. But they clearly had people on their minds too. Only when Janelle arranged for a trailer to collect the mare and foal could the barrier be breached and the foal given proper medical attention.

As the trailer bearing the mother and infant headed back to the barn, the other horses followed closely.

The foal survived, so fortunately this anecdote does not qualify as a grief story. And *this* horse circle differed in character from the quiet, still, one that formed around Storm Warning's grave. Here, the horses made a blur of motion, some moving clockwise, others counterclockwise. "Trotting, wheeling, kicking, galloping hoofed chaos," Helling recalls. She is certain that no predator, or person, could have breached that moving circle. Could the protective intent of this horse circle suggest a new possibility in relation to the geldings who surrounded Storm Warning? Perhaps they had intuited a connection between Storm and the mound that had appeared in their field, and by encircling it they meant to protect that spot and thus Storm himself. Could the horses somehow have thought that Storm might reappear? Or were they in fact mourning him?

The fact of the horse circle cannot in itself answer the question of what went on in Storm's companions' minds. But the anecdote does help to refute what some naysayers insist: that what we interpret as horse grief must instead express a feeling of vulnerability caused by separation from the herd. On this skeptical view, "grief" is an overstated claim, because the horses are only demonstrating the anxiety that besets a survivor in a herd-oriented species. Yet this "herd mentality" explanation doesn't match up with what happened after Storm died. The surviving horses placed themselves in a specific configuration and expressed no agitation through their body language. Their group was intact, save one; they had no reason to feel vulnerable. Even though we cannot intuit precisely what the horses may have been feeling, it's clear enough that something unusual was going on, beyond a concern for the self.

Responding to an article on horse grief by Kenneth Marcella in *Thoroughbred Times*, a reader described the events that unfolded after her thoroughbred filly lost her companion. This other horse, Silver, had died suddenly, and his body was visible to the filly. While Silver was buried, she was turned into a separate field. When she later returned to the field they had shared, she stationed herself on top of the grave and pawed the ground. Indifferent to offers of food and companionship, coming in at night only when forced to do so, she persisted in her behavior for almost two weeks.

Can the science of horse behavior help us understand this reaction? In his article, Marcella observes that an increase in horse longevity in the last fifteen years means that "equine buddies" now spend significantly longer periods of time together. Some horses who lose longtime friends may fall into outright depression. This is what happened with Tony and Pops, two workhorses who had known each in years past and met again at the time of their retirement. Once they rediscovered each other, these two were rarely apart. After Pops died, Tony lost weight, stopped interacting with other horses, and became lethargic enough that he lost muscle. His arthritis flared up.

In the horse world, this situation is often diagnosed as depression and treated accordingly, with anything from extra attention from human companions to doses of Valium. For horses, depression may exacerbate physical ailments such as colic, so breaking the cycle of mourning, falling sick, and becoming more depressed is potentially urgent. The introduction of a new companion may help, just as we have seen with other animals. One *Thoroughbred Times* reader told of her horse, who mourned when his pasturemate of twenty-three years died. For two weeks he stood in a spot, under a favorite tree, that he had often shared with his friend. He would not eat. Only when a mare died during foaling, and he began to care for her orphan, did his behavior turn around.

I've next to no personal experience with horses, beyond admiring their grace and intelligence—though I did, during a fourth-grade class outing, fall off a horse and still retain a memory of that long trip to the ground. Just as I remain impressed with the sheer size and power of horses, I have come to admire many horse people's embrace of horse grief and their efforts to ease it. Marcella even highlights individual variation in grief behavior, which is consonant with cutting edge animal-behavior science. As with cats, dogs, and other animals, not all horses grieve when a companion dies; the continuum of reactions ranges from extreme depression, such as I've been describing, to apparent indifference.

When a foal dies, some mares vocalize and act in an anxious manner. Others have little visible reaction. Given the strength of the mammalian mother-infant bond, I was at first surprised to learn that some horse mothers don't grieve for their foals. But on second thought, it fit with my knowledge of other animals. Through her study of chimpanzees, Jane

Goodall has expanded scientists' thinking about the variable quality of maternal behavior. Caring, competent mothers exist side by side with indifferent, neglectful mothers among our closest living relatives—indeed, within our own species—so why not in other animals too? It's possible, as well, that a mother who appears indifferent in the presence of her dead offspring may have been quite nurturing when the baby was alive, actively eliciting her care.

One pattern does seem to hold for horses, according to experts in equine behavior. "Horses given the opportunity to interact with a dead pasture mate," Marcella reports, "generally show less vocalization and anxiety and return to normal behavior more quickly." It has become a common practice to ensure that surviving horses view the body of their dead companion, in the belief that this may help them cope. What is needed for the scientific investigation of horse grief and its amelioration is a database, compiled in a consistent and rigorous manner, of reports that demonstrate a full range of outcomes, from horses who are helped by viewing a companion's body to those—like the filly who repeatedly pawed at her friend Silver's grave even after viewing his body—who are not.

Laying out the body for viewing by animals at risk for grief is an increasingly popular practice in other contexts as well. In zoos and private homes, and on farms, it's adopted by people who know that animals grieve and who want to ease that grief. It's a strategy that seemed to work with a goat named Myrtle. Myrtle knew her own mind. Adopted into a home in Colorado, she repeatedly escaped into a neighbor's yard where she could be with the only remotely goatlike animals nearby—horses. Again and again, she was brought back home, only to escape again. Finally, the wayward goat was allowed to stay where she clearly wanted to be, at the neighbor's house.

Janelle Helling, who described the horses' protective circle around a newborn foal, was the horse-owning neighbor in question. She decided that Myrtle deserved the opportunity to enjoy the companionship not only of horses but of other goats. She confesses that it wasn't a decision reached solely out of compassion for a lonely goat; there was also the matter of Myrtle's wanderlust. Whenever Janelle rode a horse off her property, Myrtle would trot behind. This wasn't safe, given local traffic.

So Janelle adopted a goat called Blondie, hoping that Myrtle might take to her and that the two goats might become homebodies together.

The plan succeeded. Four or five years older than Myrtle, Blondie was no restless wanderer. She stayed put at Janelle's. The two goats hit it off right away, and soon Myrtle began to stay home too. Janelle estimates that Myrtle and Blondie were almost always within twenty feet of one another, and often much closer. "If one of them showed up without the other," she remembers, "you knew something was wrong." A few times, one or the other would manage to get her head and horns stuck in a wire fence, requiring rescue by a human with wire cutters. Most of the time, though, Myrtle and Blondie spent the day comfortably grazing, chewing cud, playing, and napping.

Several years passed in this fashion. Then one autumn, Blondie became ill and her condition deteriorated quickly. Despite penicillin injections to combat a respiratory infection, Blondie died. This happened early on a Saturday morning, and Janelle left the goat's body unburied through the weekend as she waited to request a postmortem at the vet's. Myrtle was distressed by her friend's sudden disappearance. "Myrtle ran around the pasture vocalizing all day Saturday," Janelle says. "It was a panic-stricken scream that made your hair stand on end. She ran laps in the pasture, looking for Blondie in all their usual hangouts."

Janelle decided to arrange Blondie's body in a natural sleeping position and to make sure Myrtle could see it. That way, Myrtle wouldn't be left in the dark, as if her companion had vanished into thin air. In one sense at least, this decision paid off. Once Myrtle caught sight of Blondie, inert on the ground, her screaming and tearing around ceased. She gazed and sniffed at the body, staying with Blondie for at least twenty minutes. Myrtle then trotted away and drank some water, but immediately she returned to her old friend. Repeatedly over the next hours, Myrtle left Blondie, only to return again. Janelle interpreted this behavior as confusion, an attempt to work out why her normally active friend should be lying still. Gradually, Myrtle began to spend shorter periods with Blondie, with longer intervals between visits.

At some point, Myrtle headed out to the horse pasture. Even from there, she occasionally walked back to Blondie. By the time Monday came, though, and Blondie's body was to be taken away, Myrtle showed no interest in it. Originally, when she couldn't find her friend, Myrtle had gone into a frenzy. When Janelle oriented her toward Blondie's body, Myrtle showed keen interest; the body drew her back like a magnet. Gradually, that interest faded. Myrtle had literally moved on, away from the body; perhaps she moved on mentally as well.

Other animals may show symptoms of mourning longer than Myrtle did, even to the extent of prolonged suffering. Perhaps the mental capacities of a goat don't match those of some other mammals, but I think the more likely explanation is that this expression of grief was just Myrtle's style. Other goats might mourn differently. Myrtle reminds us, again, that grief has no singular face.

To this day, Janelle wonders if Myrtle's experience of losing Blondie was so intense in part because of her social history. As a youngster, before she was adopted by the family next door, Myrtle had been confined alone (that is, without other nonhumans) for about a year. "As social as goats are," Janelle notes, "that would have been extremely traumatic." Myrtle's first real ties were to horses. And when Blondie died, starting with the period when Myrtle transitioned away from the body, it was to horses that Myrtle went for company. Comfort knows no species bounds.

The personalities, emotions, and interior lives of farm animals—goats, pigs, cows, and domestic birds, among others—have gone largely unexplored. That situation is changing, as is beautifully demonstrated by the stories collected in Amy Hatkoff's book *The Inner World of Farm Animals*. When a cow named Debbie collapsed at the Woodstock Animal Sanctuary, other cows encircled her and bellowed so forcefully that the caretakers took notice. A veterinarian determined that Debbie's arthritis was causing her to suffer severely, and the cow was euthanized. When Debbie was buried, the other cows gathered around and vocalized with plaintive moos. Jenny Brown, the sanctuary's cofounder, observed the animals' bereavement. Not only did the cows lie down on the grave, Brown noted, "the whole group went off together somewhere on our four hundred acres and didn't come back for grains for two days. I never expected a reaction like this. I had no idea they were so aware of each other and so bonded."

Hatkoff tells also of the pigs Winnie and Buster, who had been fast friends since they were piglets at the Farm Sanctuary in Watkins Glen, New York. Five years later, Buster died. Winnie stayed to herself, refused any opportunity to interact with other pigs, and lost weight. Though healthy enough in a physical sense, she was clearly not thriving emotionally. Only when a new group of piglets arrived at the sanctuary did her mood improve. She began to run, spin, and play with the piglets and slept with them at night, all behaviors reminiscent of her patterns with Buster, who by then had been gone for two years.

Last year, I adopted a chicken at Farm Sanctuary. Fiesta is a striking black hen who was found wandering a park in the Bronx. Her rescuers think she may have escaped from a Santeria ritual involving animal sacrifice, purportedly the source of dead chickens found previously in the neighborhood. Whether that was the case or not, the homeless hen was brought to safety. My adoption of her doesn't mean that Fiesta now strides around my backyard, dodging our cats; it's just that I help pay for her care at the Watkins Glen sanctuary.

With two more facilities in California and a significant national presence, Farm Sanctuary protects farm animals and urges people to think about them in fresh ways. Farm animals are "someone, not something," as a recent campaign puts it. "We can tell you from personal experience," staff members write on the Farm Sanctuary website, "that farm animals have the same range of personalities and interests as cats and dogs."

As we have already seen, an animal who is *someone* may love and may grieve. In 2006 three mulard ducks were rescued from a foie gras farm and brought to Farm Sanctuary. Foie gras, literally "fatty liver," is a food product made by force-feeding ducks and geese, a practice that causes the animals to suffer. All three mulards showed signs of liver disease called hepatic lipidosis. The two in the worst shape were males called Harper and Kohl. Because of fractures that went untreated at the foie gras farm, Kohl's legs were deformed. Harper was blind in one eye. Both ducks were terribly frightened of people. The single blessing in the whole situation was that they became close friends and chose to spend almost all their time together.

That Kohl and Harper lived for four years at the sanctuary was, given their traumatic past histories, a happy and unexpected outcome. When Kohl could no longer walk, or his pain be treated effectively, he was euthanized. From outside the barn where the procedure took place, Harper was watching, and after it was over, he could see his friend's body, lying in straw on the barn floor. At first, Harper tried to communicate with Kohl in the usual ways. Getting no response, he bent down and prodded Kohl with his head. After more inspection and prodding, Harper lay down next to Kohl and put his head and neck over Kohl's neck. He stayed in that position for some hours.

Harper got up eventually, and sanctuary caretakers removed Kohl's body. For a while after that, Harper went every day to his favorite spot, once shared with Kohl, next to a small pond. There he would sit. Efforts to introduce him to another potential duck friend didn't take, which was especially sad because Harper was now more nervous around people without Kohl. Everyone at the sanctuary recognized Harper's depression. Two months later, Harper died as well.

Harper and Kohl could be the poster ducks for the book's theme: Where there is grief, there was love.

4 WHY BUNNIES GET DEPRESSED

Over the years, my family has cared for two rescued rabbits: one large caramel-colored, long-haired angora male, and one petite Oreo-colored, short-haired female. In a burst of creativity, we named these rabbits Caramel and Oreo.

Caramel had been a classroom bunny at a Montessori school attended by my daughter. At the school, the children enjoyed the freedom to move around and explore their schoolroom as they learned, but the class rabbit endured confinement. Allowed out of his cage only infrequently, he badly needed more space to hop around. With the blessings of the school, we adopted Caramel and gave him that space. In our house, Caramel lived to age eight, evidently enjoying our company, if only tolerating that of our cats (who tolerated him back). When Caramel died, we adopted Oreo from the animal shelter, and a similar happy trajectory ensued until she, too, died of natural causes.

Because these two bunnies never met, I had no opportunity to witness rabbit friendships, or any kind of rabbit-to-rabbit interaction. Certainly I'd noticed how affectionate Caramel and Oreo could be with us, when the mood struck: they would seek us out, and push their noses against us or relax their bodies into our caressing touch. Caramel joined the family in front of the TV, hopping from his turf, the back of the house, into the den to recline on a throw rug we laid out for that purpose. Oreo preferred to leap onto a couch and sit right beside me.

I do peek in occasionally on Jeremy and Jilly, two rabbits who live around the corner from me. Jeremy, a Tennessee Red, was adopted first by my friends and fellow cat rescuers Nuala Galbari and David Justis. Once saved from a local pet store's small cage and settled into a home full of loving care for animals, including cats and birds, Jeremy thrived. Like Janelle with the goat Myrtle, though, Nuala and David felt that same-species companionship was in order. Enter Jilly, an older rex female. Despite a whopping (for rabbits) six-year difference in their ages, Jeremy and Jilly bonded quickly. Around and around the bedroom rug they now race, with play leaps and twists into the air. In quieter moments, they groom each other. When restricted to their enclosure, at night and for certain periods during the day, they have ample room to spread out but still often press their bodies tightly together.

Jilly's agility and verve mask her age, but she is nine years old now. Watching Jeremy delight in her company, it's natural to wonder what response he may have if Jilly, as seems likely, dies first. Michelle Neely shares the story of rabbit companions Lucy and Vincent. Michelle and her husband adopted Vincent from a animal shelter in pretty bad shape; abandoned by his previous owners, he had been half-starved, with mites and scabies on his body. For six months, he was Michelle's only rabbit, and a fiercely affection-seeking one at that. Vincent loved to be cradled like a baby, in Michelle's or her husband's arms or lap. Content to stay still for periods of thirty minutes to an hour, he received a bunny massage from his human friends.

Next the couple adopted Lucy. Like her two brothers, Lucy had been born without ears. Where those iconic floppy bunny ears usually are, Lucy sported only cartilage nubs. She was totally deaf. Yet Lucy had been around other bunnies for prolonged periods, and she knew how to act socially. Not so Vincent. For three months after introducing the two, Michelle worked with them every day, coaxing them to bond. It went slowly, because Vincent didn't know how to signal to Lucy that he wanted to groom or otherwise be friendly. What started out as a promising, fun interaction often veered into mild aggression between the two rabbits. Whether Lucy's earlessness contributed to this state of affairs isn't clear. Maybe some of the back-and-forth signaling between them was compromised by Lucy's unusual anatomy. Possibly her un-rabbitlike appearance affected Vincent's responses in some way. Whatever the factors in play, this was no love-at-first-sight scenario.

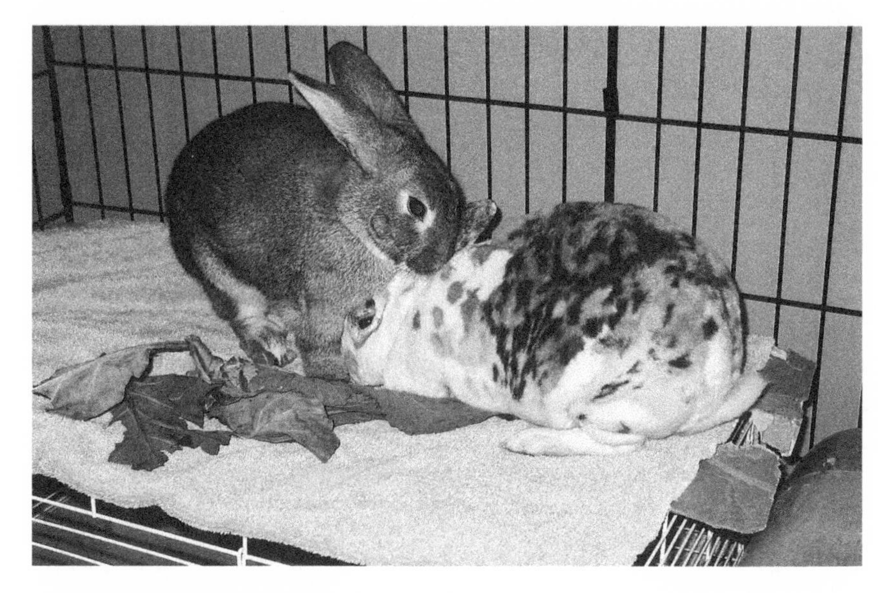

JEREMY AND JILLY. *PHOTO BY DAVID L. JUSTICE, MD.*

Then, out of the blue, Lucy jumped into Vincent's pen and spent the night. When Michelle saw them together the next morning, she recognized that a dramatic transformation had taken place: Vincent and Lucy were bonded. Really bonded, to the extent that Vincent seemed almost lovesick for Lucy. Just as I've watched Jeremy and Jilly do, the two would race around in the morning, playing and wearing themselves out, then sleep together later in the day. "Lucy was always the leader of their little expeditions," Michelle explains, "up the stairs, or around the living room, or onto the balcony. Vincent followed her everywhere, because he always wanted to be near her. Watching him with Lucy, you almost thought that he hadn't known there were other rabbits in the world, and now that he had discovered it, he was lost in the wonder and sheer delight of that fact."

Sadly, Vincent and Lucy shared only eight or nine months together. Then Lucy got sick, with incurable infections in both ear canals that probably stemmed from her congenital condition. Despite surgery by an experienced veterinary team, she died. Vincent, Michelle says, "spent about a week doing tragic sweeps of the house, searching for her." After that, he seemed to grasp that Lucy was not coming back. He fell into the sort of depressive state that will by now sound familiar: he stopped eating much and refused to leave his "rabbit condo." Inside that house, he sat in Lucy's preferred spot and did little else. The vigor he had shown when playing with Lucy was completely absent.

Michelle began to fear that Vincent too would die. She adopted a new rabbit, Annabel, hoping that Vincent might perk up. This he did; immediately upon meeting Annabel, his interest in everyday activities revived, as did his appetite for food.

Now, this serial bonding—first to Lucy, second to Annabel—might raise some questions. Could it be that Vincent wanted another rabbit nearby simply because he didn't care for solitude? Did he care one way or the other whether that other warm bunny body belonged to Lucy or Annabel or someone else? Had he forgotten all about Lucy?

Since we can't know Vincent's thoughts, we can attack these questions only by taking a close look at the events in the months after Vincent met Annabel. He acted differently than he ever had before. At even the briefest separation from Annabel—say, when Annabel tucked herself into a corner of the apartment to nap—Vincent became anxious. He would search all over for her, with increasing distress if the search was unsuccessful. "Finally," Michelle says, "we would pick him up and take him to wherever Annabel was, at which point he would relax immediately." It seemed to her that Vincent feared losing Annabel as he had lost Lucy.

Seven months into his new friendship, Vincent's anxious behavior stopped. Whether he came to trust the fact that Annabel wouldn't disappear, whether he'd forgotten Lucy, or whether his behavioral shift owed to some other factor, no one knows. When dealing with rabbits, we shouldn't assume that quick bonding to a new partner implies a lack of genuine mourning for the lost one, any more than we would assume that for bigger-brained mammals, including ourselves. In fact, why not turn our thoughts 180 degrees: Could it have been the deep satisfaction that Vincent experienced with Lucy that led him to revive so quickly when Annabel came on the scene? Perhaps the sight and smell of Annabel gave Vincent the rabbit equivalent of hope for renewed companionship. On the other hand, Michelle observes that Vincent and Annabel developed a quick-and-easy friendship but one that was less intense than Vincent and Lucy's, even despite Vincent's anxious searching early on.

After Michelle first contacted me, we began an e-mail correspondence. Several weeks into this exchange, Vincent died. This time, Michelle did something new. She showed Vincent's body to Annabel. Annabel sniffed and licked her friend's still form, then went away from and back to the body, the same pattern followed by the goat Myrtle when Blondie died. When Annabel tried to move Vincent's body out of the condo she had shared with him, Michelle took it away for cremation.

Annabel didn't seem to mourn for Vincent as Vincent had mourned for Lucy, over a period of weeks. Here we see evidence of two relationships that differed, and two survivors whose arcs of response to a loss differed as well.

The House Rabbit Society (HRS) is an animal-rescue organization headquartered in California but with an international reach. Its mission is to rescue abandoned rabbits and educate people about proper rabbit care. Its website is crammed with links to explore, including "Just for Fun: Rabbits and Their Sense of Humor," "Living with an Aloof Rabbit," and "Understanding the Emotional Messages of Your Rabbit." These experts embrace the notion of rabbit grief. They would not hesitate to conclude that Vincent mourned Lucy.

The HRS offers grief stories of its own, and these underscore the fact that rabbits' responses to death vary greatly. Some rabbits, I learned, exhibit an unusual behavior: If they are present when a cagemate or close friend dies, they leap into the air in a kind of dance. I've seen no explanation for this action, although it is described as a sudden release of energy.

Other rabbits may "act out" and misbehave. Upon losing his companion Dinah, a four-year-old rabbit named Lefty continued to act in his usual high-spirited manner. No echoes of grieving Vincent here. Instead Lefty jumped up onto the bed "his people" sleep in and chewed holes in the pillowcases. The HRS cautions that, in this context, a cheeky rabbit may need extra TLC and perhaps a new same-species friend, because grief may present itself as misbehavior.

Through the HRS, Joy Gioia tells of grief in a bunny trio. Just as Vincent, Lucy, and Annabel formed a sort of emotional triangle with Vincent at its center, responding first to Lucy and later to Annabel, so it was with Trixie and her two successive companions, Joey and Majic. In this case, two of the three rabbits had fared quite badly as people's pets. All three were rescued by Joy, a volunteer rabbit fosterer associated with the HRS.

The story starts with Joey. Because his original caretakers were neglectful, he had suffered from severe infections that left him totally blind in one eye and partially blind in the other, which leaked liquid constantly. He was also deaf and saddled with breathing problems. Emotionally, he had more or less shut down; he especially hated the cleaning that his bad eye required and so was no fan of interaction with Joy or any other human. It would have been the easy choice for his human caretakers to euthanize Joey, but that is not what the HRS is all about.

Trixie had not been so poorly treated, but she needed surgery to remove her incisors because of severe malocclusion. Like Joey, she wasn't keen on interacting with humans. By chance, she ended up in foster care with Joy, housed right next to Joey. Each rabbit showed sparks of interest in the other, and to encourage these, the pair was moved into a larger, sharable enclosure. This matchmaking plan worked spectacularly well. Trixie tended to Joey with devotion, including cleaning his bad eye. She licked it gently, which Joey definitely preferred to having it washed by humans. A great deal of affection was evident between the two.

A third bunny came into foster care also, but this one preferred people and wanted nothing to do with other rabbits. Majic had been a classroom bunny for five years. That might sound like a fairly good life for a rabbit, but as we discovered ourselves when we adopted Caramel from the Montessori school, even well-meaning schools can offer inadequate resources. Majic's cage was small, too tiny even for him to properly clean his ears. It had a wire floor, which is hard on rabbits' feet, which lack the thick pads that cats' and dogs' feet have. And having excitable children surrounding one's cage can't be the most restful experience. Eventually, Majic began to lash out at children who put their fingers into his cage.

When he arrived at foster care, Majic suffered from a bad ear infection, molars in such poor shape that he couldn't eat properly, and nerve damage in his feet from a declawing procedure. Curing Majic's ear and dental problems wasn't all that hard. The nerve damage, though, meant he couldn't jump well. He began to relax around people and even to enjoy cuddling, but he acted defensively when put with other rabbits. He was kept isolated, in a house with thick, comfortable sleeping material, in the same room as Trixie and Joey. A perimeter was set up around his bed to make sure the other rabbits kept out, but no latched gate was deemed necessary because Majic made no effort to get out on his own.

For two years, Joey and Trixie enjoyed their friendship. Then, Joey began to lose weight and his health worsened. He suffered a seizure. Joy and her husband consulted with a veterinarian, and they all agreed it was time to let Joey go. He was euthanized, and at the vet's, Trixie was allowed to stay with his body for a while.

Back home with Joy, Trixie was sad. She didn't eat and, Joy recalls, she "made a small and pathetic picture as she lay in her empty house." The next morning, though, Joy viewed an unexpected scene: Majic had jumped down from his bed and the two rabbits lay close to each other, prevented from close contact only by the door of Trixie's house.

After settling Majic back on his bed (for the sake of his damaged feet), Joy opened up a channel between their houses. For two days, Trixie shuttled back and forth between the two "home turfs," and then she moved in with Majic. By the third day, the new friends were cuddling and grooming, and Trixie once again was eating well.

Trixie was one of the lucky ones, just as Vincent had been. Some rabbits don't come out of it so easily. Rabbits, like many other animals, may fall into serious depression when they mourn. In extreme instances, they may even starve themselves to death.

Just as I was exploring the nature of severe depressive responses to grief, Karen Wager-Smith sent me a paper on the neurobiology of depression that she wrote with Athina Markou. Focusing on a wide variety of animals, including humans, Wager-Smith and Markou ask whether an understanding of the dynamic brain might cue us in to adaptive aspects of acute depression, the type that's both symptomatically intense and relatively short in duration. The two scientists posit a chain of events that culminate in a person's experience of acute depression. The trigger is some kind of stressful life event. Perhaps a person loses her job or faces an unwanted divorce. Or he may be sent for repeated tours of combat or lose his partner to death. Studies show that about three-quarters of initial depressive episodes are preceded by major stress of this kind.

What happens next occurs at the neurophysiological level. The dynamic nature of the brain means it's time to jettison old assumptions about an organ that remains fixed and static in adulthood, after a period of growth and adaptation during one's younger years. In fact, our brains always grow and adapt at the physiological level. Each of us sees, thinks, and feels our way in response to events that occur (or that we create), and as we do, our brains rewire. In tandem with our experiences, neurons are strengthened or fall away. Some of the neural pruning that goes on, we might initially think of as negative: a loss of brain tissue, after all, doesn't sound welcome. That's where the second step in Wager-Smith and Markou's sequence comes in.

Wager-Smith and Markou describe types of "microdamage" that stress may inflict upon the brain and that reduce key neuronal connections in certain regions. Data from animal models suggest that in two brain areas, the hippocampus and the prefrontal cortex, synaptic material may be reduced in the aftermath of stress. Because the hippocampus deals with memory and emotion, and the prefrontal cortex is a center of planning and personality, it's clear that such damage, even if limited, could affect an animal's perception of the world. And it's not only that animal models predict changes in people's brains because of stress. Recent studies in brain-imaging suggest a causal relationship between long-term depression and shrinkage of certain brain regions in people, most clearly the hippocampus.

But just as our body rushes to respond when we experience trauma to a limb or infection in an organ, the brain acts to protect itself from the insults of stress. The next step in the chain is brain repair, kicked off when the micro-damage triggers an inflammatory response. And just as there may be short-term negative consequences during recovery from trauma or sickness, in the period after the brain-shock of stress, a person may feel fatigued, sleep more, and eat less. When the brain is involved, there's the likelihood of something extra, too—a tendency to feel acute emotional pain. Wager-Smith and Markou think that the inflammatory response may cause a kind of hypersensitivity to psychological pain.

In many cases, this hypersensitivity is of limited duration; as the repair mechanisms do their job, the mental anguish begins to fade. Unfortunately, though, the pain doesn't always diminish. For some people, acute depression, a response to a stressful event, settles in, in a soul-crushing way. Systems as complex as the human brain are prone to highly variable outcomes depending on a web of factors—from genetic predispositions to family patterns, from personality traits to access to resources that might bolster one's ability to cope. If for some reason the hypersensitivity to mental pain becomes entrenched, a person may fall into the "unrelenting pain" of extended depression that William Styron describes in his memoir *Darkness Visible*.

The synopsis I've just provided only summarizes a detailed hypothesis. In their article, Wager-Smith and Markou present neurobiological evidence to support each step in this proposed chain. Because its explanatory power is rooted in both biology and culture, this model of human adaptation is of a type that anthropologists like me admire. It takes into account lived experience as much as aspects of physiology and genetics. Again, it's not just that the brain shapes our responses to what happens around us, but that what happens around us sculpts our brain, and continues to do so throughout our lives.

Within limits, then, a depressive response to severe stress may be beneficial. When a person or other animal is shocked by a life event, it may benefit that individual for the brain to stage a mini shut-down. The sufferer thus gains time to recover emotionally. The new neuronal connections may, in Wager-Smith's phrase, "mediate new behavioral strategies" for the individual as she tries to move past the stressful event.

This model is a significant addition to previous theories, concisely reviewed by John Archer in his book *The Nature of Grief*. Archer notes that in evolutionary terms, grief may be maladaptive, compromising an animal's capacity for survival and reproduction. A grief response may be a sort of exacerbated separation response. The separation response, which occurs when two animals who matter to each other find themselves apart for some reason, involves distress, protest, and behaviors directed toward reuniting with the lost partner. As such, it may increase the chances of reunion and thus be adaptive. In some cases, a grief reaction may ensure that a partner in a separated pair doesn't too quickly turn to a new mate, when the missing animal may yet return. In other cases, there is no apparent benefit; grief then may be just a natural, highly elaborated by-product of the separation response or, more broadly, of the animals' close bonds.

Archer does discuss depression as it relates to grief, but Wager-Smith and Markou's model goes further, explaining more precisely why it may not be pathological for some animals to exhibit *severe* grief at the death of a friend or partner. If stress has rewired an animal's brain, a period of altered sleeping and eating may conserve energy in a way that aids psychological as well as physical healing. The sadness—indeed, mental agony in some cases—is the "extra" that comes with this brain stress. As Wager-Smith told me, "Grief is an evolved behavioral program, akin to sickness behavior, that promotes convalescence during a significant neural rewriting job."

Perhaps, when a survivor pairs up with a new partner, the brain-repair process accelerates, allowing for quicker recovery. We have seen, with the rabbits Vincent and Trixie and with other animals, that a new social stimulus may snap an animal out of lethargy. In proposing a causal link between acquiring a new partner and a "kick" to brain recovery, I'm only speculating. Wager-Smith and Markou's model itself may prove right or wrong in its details, or even in its major points. It's the way of science, and an elegant one at that, to propose some explanation, multistepped and intricate, that must then be tested, by its authors and others, as they gather more data.

Surely there is no single, overarching explanation for all episodes of depression in people or other animals. Yet the beauty of Wager-Smith and Markou's model lies in a reminder it offers: Because death and mourning surely count as one of life's most stressful events, there may be a common biological underpinning to the grief that animals—horses, goats, rabbits, cats, dogs, elephants, chimpanzees, and people—feel. To make this suggestion is not to say that we are hard-wired creatures whose brains all respond in identical ways. It is, rather, to take seriously the notion that we mammals share some tendencies in our biology and in the ways our life experiences may affect our biology. Though based on that common platform, outcomes will—because of species-specific behaviors, different developmental histories, and individual personalities in complex combination—be variable, both across and within species.

Meanwhile, I like to think of rabbits as "iceberg" animals in the world of animal grief. Bunnies—like the chickens and the goats I discussed in the prologue—are not the animals people first think of when considering nonhuman grief. They are the tip of the iceberg because they point us toward a future time, maybe not too long from now, when the fact of animal mourning across diverse species will be taken as common knowledge.

5 ELEPHANT BONES

When elephants grieve, the emotion may stream from those huge, wrinkled-gray bodies in palpable waves. If you are close enough, you can feel it in the air.

Animal-behavior expert Marc Bekoff went to northern Kenya with Iain Douglas-Hamilton, one of the world's top elephant scientists, and was startled when he first spied the massive creatures. "Their heads were down," Bekoff reported, "ears dropping, tails hanging listlessly, and they were just walking here and there, moping around, apparently broken-hearted." He first felt the elephants' emotion, then learned from Douglas-Hamilton that the herd's matriarch had recently died.

As the two scientists continued their drive, they came upon a second group of elephants only a few kilometers away. Here the scene was much different. These elephants looked content. With heads up, ears up, tails up too, they exuded a sense of well-being.

That the first, sad group was in mourning—and not just a bit disorganized after the loss of their leader, or momentarily upset about some other matter—we can pretty much assert as a fact. Example after example of mourning by elephants who have lost one of their tightly bonded group has been reported by scientists. It's the closest thing we have, in the nascent world of animal-grief study, to scientific certainty. In this way, elephants are a touchstone species for understanding how wild animals grieve.

Douglas-Hamilton's own years of elephant study prove the point. Since 1997 his research team has monitored the population at Samburu

National Reserve in Kenya, where nine hundred elephants are known as distinct individuals. (This feat is impressive. At Amboseli, in southern Kenya, it took real effort for me to learn reliably the identities of just over a hundred baboons.) A year later, GPS technology was added to the scientists' arsenal, so that radio-tracking data now supplements direct observation of the elephants.

At Samburu, as in other elephant populations, the females and their young offspring form tight units; female kin and favored associates tend to stay together, or break up into smaller units that reunite into a herd at regular intervals. Most of the year, the mature bulls roam independently, approaching the herd only to mate with fertile females.

Judging from a fantastic discovery in 2011, prehistoric elephants organized themselves in just this way too. Distributed over a large area of desert in the United Arab Emirates, and studied in part from the air because of its sheer size, is an ancient "trackway" of elephant footprints. These footprints, which at first glance appear to be mere depressions in the earth, tell us that at least thirteen elephants of varying sizes and ages walked together seven million years ago. Separately, there walked a much bigger elephant. If, as scientists suspect, this separate animal was a solitary bull, the footprints amount to a prehistoric blueprint for the social organization of elephants today.

The geometry of the dual fossilized paths is telling. The narrow trajectory of the footprints of the thirteen animals suggests that these elephants moved in concert. The lone elephant's footprints transect those of the herd, meaning that he was moving in an almost perpendicular direction. Paleontologist Faysal Bibi described the footprints to the BBC as a "beautiful snapshot" of the social behavior of a now-extinct ancestor of today's elephants.

Long-term Amboseli elephant researcher Cynthia Moss depicts the relationship scheme of modern elephants as a concentric circle: females and their young occupy the center. In the next ring are other female relatives, like sisters and grandmothers. In the outer rings are the males, first the younger ones, hovering on the verge of independence, and finally the mature roaming bulls.

When family members who have been separated by their divergent travels come together again, their reunion is a virtual choreography of joy. The elephants intertwine their floppy trunks, click their tusks together, and flap their ears. Out may gush fountains of urine. Sometimes the bulky bodies spin around and individuals back into each other. Throughout this period, which may last as long as ten minutes, there's vocal accompaniment: rumbles, screams, and trumpets.

On joy's flip side, we enter the territory of grief. Douglas-Hamilton's team recorded a startling event in 2003 centered on the elephant Eleanor, matriarch of a family called the First Ladies. Well-known to the research team, Eleanor had been seen 106 times over the years. About five and a half months before the incident I'm about to describe, she had given birth. Her calf, a female, was doing fine. Maya, an elephant who had been seen 101 times, was Eleanor's closest companion, and the researchers strongly suspected that she was Eleanor's daughter.

On the early evening of October 10, Eleanor was seen dragging her swollen trunk on the ground. One ear and one leg appeared bruised. As Douglas-Hamilton and his coworkers later reported, she took a "few slow small steps" and then was seen to be "falling heavily to the ground." Two minutes later, Grace, matriarch of a family called the Virtues, approached. Using her trunk and one foot, she explored Eleanor's body. Grace then used her tusks to lift Eleanor back onto her feet. Too weak to stay upright, Eleanor collapsed again when Grace pushed her and urged her to walk.

Grace must have understood something of Eleanor's dire physical state, because she exhibited extreme distress as she came to Eleanor's aid, vocalizing and continuing to push Eleanor with her tusks. Even when the rest of her family moved on, Grace stayed put for at least another hour, right at Eleanor's side. At this point, Maya, the presumed daughter, was far away and could have had no idea that Eleanor was down. In fact, Eleanor never got to her feet again. She died the next morning.

From radio-tracking we know that on the second day, Maya came within ten meters of Eleanor. But it was an elephant called Maui, from a family known as the Hawaiian Islands, who showed a keen response to Eleanor's body. She extended her trunk, sniffed and touched the body, then put her trunk into her mouth to taste it. With her right foot, Maui hovered over Eleanor; she nudged her and pulled the body with her left foot and trunk. I assume she was trying to right Eleanor, as Grace had the day before. Then Maui did something different: she stood over the body and rocked back and forth. Together these actions lasted for eight minutes.

For a full week after Eleanor's death, elephants came to the body in a parade of exploration and emotion. On day three, the park rangers cut the tusks from Eleanor's body, an act that must have been designed to thwart ivory poachers. From that point forward, what remained was a severely disfigured elephant body. Eleanor's trunk was severed, and where the tusks normally would be were only open holes.

On this same day, Grace returned to Eleanor. This time, she made no move to lift the fallen elephant; she only stood quietly by the body. Maya and other members of Eleanor's family came near. They didn't, as far as I can tell from the reports, touch their matriarch's body—with one exception. Eleanor's young daughter, the new calf, nuzzled her mother. She seemed disoriented, trying to suck from other young calves, then returning to the body of her mother.

Ultimately, this calf was not to survive. Though she was seen in subsequent weeks attempting to suckle from other breeding females in her mother's group, none of those elephants obliged her—and she was too young and vulnerable to survive without milk. But on day three, the calf seemed only to want to be near her lifeless mother. When an unrelated family named Biblical Towns approached, its members pushed away the First Ladies—that is, Eleanor's kin—in what appeared to be a combination of dominance bid and desire to explore the dead body. Only Eleanor's calf was not pushed aside. A photograph shows her standing next to her mother, alone, near to a group of large and imposing elephants who were not her kin. The baby's stiff posture, and slightly extended trunk, makes a poignant image.

Over the next four days, Maya and other members of the First Ladies family spent some time near Eleanor's body and some time away. By the fourth day, the carcass had become a scavengers' feast: jackals, hyenas, vultures, and lions all ate from it. On the sixth day, a female called Sage, from the family Spice Girls, approached the body. Even at this point—

with her tusks missing and her carcass partially consumed—Eleanor evoked a response. Sage spent three minutes sniffing and touching the body with her trunk.

At no time during the week following Eleanor's death did a bull visit the carcass. The responses came from females, but not only from Eleanor's female kin. Five elephant families demonstrated a distinct interest in the body, including Eleanor's own family. In their report, Douglas-Hamilton and his coauthors find it significant that elephants' keen interest in dying and dead individuals is not limited by genetic relationship. "Elephants have a generalized response to suffering and death," they conclude.

The study of elephants' responses to Eleanor's death lasted one week, but elephants almost surely remember their dead for much longer than that. If at Samburu, radio-tracking data supplement what we learn from observational work, at Amboseli, experimental trials that measure elephants' responses offer a fresh perspective on how elephants respond to the dead.

I admit to a special fondness for the Amboseli elephant research. For one thing, there are twenty-two hundred elephants individually known at Amboseli—if I found the nine hundred at Samburu impressive, let's just say I'm knocked out by this higher figure and the intensive labor by elephant scientists that it represents. Also, Amboseli is where I spent fourteen months baboon-watching, and where I had the incomparable experience of observing elephants who lumbered right into my backyard. Moss's elephants (as I think of them) ventured surprisingly close to the Amboseli Baboon Project's thatched adobe house, where I lived. At night, through my bedroom's open-mesh window, I heard the push of their great bodies moving slowly through the vegetation. During the day, I viewed them silhouetted against snow-laden Mount Kilimanjaro, which loomed large across the Tanzanian border. Though my beat is primates and I have never formally studied the Amboseli elephants, my casual encounters with them at my house and in the field were unforgettable.

The idea that Amboseli elephants seek out the bones of their deceased loved ones to caress seemed marvelous to me; it wrapped up elephant smartness and elephant emotion into one package. I reported to others that elephants distinguish the bones of their dead relatives from those of other elephants in their habitat, and behave differently toward the bones of their kin. It's not that this information is downright false or mythical. It's not on a par with the popular but apocryphal idea of an elephant graveyard. (Elephants do not travel purposefully to a single place to die. They may travel toward water and end up dying in clusters near water with more than random regularity, or they may, sadly, be shot by humans in such numbers that their strewn carcasses resemble a graveyard.) In fact, I got the idea that elephants seek out the bones of dead relatives from Cynthia Moss herself.

In her book *Elephant Memories*, Moss tells a story of bringing back to camp the jaw of an elephant matriarch who had died a few weeks earlier. Three days later, the elephant's family passed near the camp. When they smelled the jaw, they diverted their course to approach it. When the family finished its inspection and moved on, one elephant stayed behind. The dead female's seven-year-old son continued to stroke the jaw and turn it with his foot and trunk. Moss felt certain that the young male somehow recognized his mother.

Other Amboseli elephants' response to the bones of a relative—their matriarch—was captured on film. A small group of elephants encircles the bones on the ground. Some elephants begin to turn the bones over and pick them up in their trunks, feeling their nooks, crannies, and crevices. It's the detailed exploration of the bones that is so striking—all while the elephants, at least some of them, vocalize. Then, with the bones on the ground again, the elephants touch them with their back feet.

It's common enough for Amboseli elephants to caress bleached bones that they come across in their travels. But does the exploration of the bones recorded in this film (and that described by other elephant researchers) equate to grief, as the narrator suggests? Might the strength of an elephant's response to bleached-white bones somehow correlate with her degree of kinship with the dead elephant? It sounds plausible, since we know that elephant relatives bond deeply with each other, that elephants have enduring memories, and that elephants grieve. Is it so strange to think that elephants are able to identify the bones of loved ones who have died some time before and visit them in order to pay homage?

Karen McComb, Lucy Baker, and Cynthia Moss have attacked these questions experimentally at Amboseli, in a superb example of how science works. The three scientists set about rigorously following up on an impression Moss had gained from her casual observations of relatives' responses to the bones of the dead. The research questions were these: Are elephants more attracted to skulls and ivory from elephants than to other objects? Do they show more interest in elephant skulls than in skulls of other large mammals? Do they prefer to investigate the skulls of relatives over the skulls of other elephants? Yes, yes, and no are the answers, according to the experimental data. Elephants care very much about their own species' bones compared to other objects or the bones of other species, but show no evidence of preferring the skulls of their own kin to the skulls of other elephants.

First, McComb and her coworkers presented a piece of ivory, a piece of wood, and an elephant skull to an array of elephant families (one family at a time). The arrangement of the objects was carefully controlled, with different objects in the rightmost, center, and leftmost positions from trial to trial. The elephants' behavioral responses were videotaped, with particular attention given in the analysis phase to the length of time an elephant spent exploring an object with trunk or feet. Of the three objects, elephants preferred the ivory. The skull was next, and the wood came in last. Since the skull is, of course, invisible during an animal's life, I wonder if the ivory was preferred because the elephants more readily recognized it as belonging to a particular individual, perhaps through a scar, chip, or discolored area. The researchers hint at this possibility by noting the ivory's connection with living elephants.

Next, three skulls, one each from an elephant, a buffalo, and a rhino, were arrayed in front of elephant families. The elephants distinctly preferred the skull of their own kind, with a lesser but equal attraction to skulls from the other two species. The third part of the research involved three elephant families, each of which had lost its matriarch in the past one to five years. The surviving elephants were presented with the skulls of the three matriarchs, only one of whom, of course, had been their own group's matriarch. The elephants showed no greater interest in the skull of their own matriarch.

What, then, is the meaning of the anecdote Moss reported, about the seven-year-old son caressing his mother's bones? Do the experimental results negate the emotion the son seemed to express as he lingered over the remains of his mother, or the suggestion that elephants more generally may mourn their loved ones by caressing the bones? I think the answer to that last question is no. When anecdotes are reported by scientists or others who cautiously interpret the behavior of animals they know well, they point to an animal's capacity to carry out some action or express some emotion. Even if only some elephants grieve for a lost relative or friend, even if only some elephants caress the bones of their dead relatives, that behavior is genuine, and meaningful, for those individuals.

When it comes to animal emotion, today's animal-behavior science tacks back and forth between analysis of events noted by credible observers and evidence derived from controlled experiments (which, as the Amboseli researchers show us, can take place in the field as well as in captivity). These two sources are complementary. The events reported may be rare but hint at unsuspected possibilities and emotional depth in the animals' behavior; the controlled evidence requires us to put the brakes on reckless speculation about those alluring possibilities. The experiment by McComb, Baker, and Moss constrains wild statements to the effect that elephants (implying all elephants) recognize and prefer the bones of their dead relatives, and this in turn constrains speculation about the way elephants (implying all elephants) mourn.

The McComb study tells us that elephants are keenly intrigued by the bones of their own kind. In day-to-day life, this tendency surely means that they are attracted to, and that they interact with, bones of their kin (as well as bones of non-kin). How readily elephants recognize their own kins' bones and mourn the individuals represented by those bones remains a mystery.

Once, the Echo family of elephants in Amboseli came upon a carcass of a young female who had been sick for a number of weeks. That the elephants explored the carcass comes as no surprise, but they went on to do something remarkable. As Moss watched, the elephants

> began to kick at the ground around the carcass, digging up the dirt and putting it on the body. A few others broke off branches and palm fronds

> and brought them back and placed them on the carcass. At that point the warden circled overhead and dived down in his plane to guide the rangers on the ground to the dead elephant so that they could recover the tusks. The [elephants] were frightened by the plane and ran off. I think if they had not been disturbed they would have nearly buried the body.

A near-burial by elephants must have been remarkable to witness. Was this interrupted act an attempt to protect the deceased elephant's body from harm? Why have other long-term elephant researchers not reported (at least to my knowledge) such behavior? Elephant burial of a body cannot be a common act; too many field hours by scientists have accumulated for such a behavior to have been missed were it routine. But it's impossible to dismiss Moss's account, given her intimate knowledge of these elephants.

At the Elephant Sanctuary in Hohenwald, Tennessee, elephants are also known intimately, in this case by their caretakers. Elephants' past histories (mostly in the entertainment or zoo world), how they are adjusting to their new life, with whom they make friends, and how they express their personalities are all noted. Knowledgeable and caring eyes witness subtleties of elephant behavior, then share these with the larger elephant-loving community by posting them on the sanctuary's website, which has a section devoted to each elephant resident. I have developed a special fondness for the elephant Tina's story.

Tina was born in 1970 at a zoo in Portland, Oregon. At two years of age, she was sold to a game farm in British Columbia. For fourteen years, she lived alone in a barn, accompanied only by a Saint Bernard dog named Susie. Once in a while, she enjoyed the overnight visits of the owner's children to the barn. How long must those fourteen years have felt to Tina, who endured so much solitude day by day and hour by hour? Finally, Tina was joined by another elephant, Tumpe. These two females were allowed to remain together even after the farm was sold and turned into the Greater Vancouver Zoo. Again a zoo resident, Tina at least had another elephant for company—until 2002, when Tumpe was sold to yet another zoo, this one in the United States. Tina was alone again.

At this point, Tina wasn't in the best of health. She weighed too much, and her feet gave her problems, conditions that often beset captive elephants. The Canadian zoo staff not only cared for Tina, they cared about her, enough that they chose to release her from the severe physical and emotional constraints of life in the zoo. In August 2003, Tina was transported three thousand miles to the Elephant Sanctuary in Tennessee. There she discovered what she had so long been forced to live without: the sustaining company of others of her own kind.

This happy outcome didn't come easily. It required patient coaxing from emotional coaches of two species: human and elephant. After all, Tina had not been around more than one elephant at a time; suddenly she had to contend with a host of incoming social signals and to negotiate a web of elephant relationships. By early 2004, Tina still was hesitant in some of her social interactions. When more than one elephant would enter the group stall, out she would go.

One night in mid-January, first the elephant Tarra and then another named Jenny came into the stall and began to rub up against Tina. Though Tina moved into the next stall, she elected to remain near both females. When Jenny barged right on in to Tina's stall, Tina acted possessive toward her ball and her hay, but she eventually relaxed. This was a step forward. That same month, a bond bloomed between Tina and Winkie. The caretakers noticed that Winkie seemed to want a secretive social tie with Tina. It had taken Winkie over two years to integrate into the sanctuary's herd. Now, she seemed to crave affection from Tina, yet at the same time hid evidence of it from human eyes.

This behavior is understandable in light of Winkie's own history. Wild-born in Burma, she was captured at the age of one and transported to a US zoo, where the staff managed her behavior with harsh displays of dominance. It took Winkie years to unlearn the toughness she assumed at the zoo, but it did happen; when she began to stand close to and touch Tina, her gentleness was encouraged by the sanctuary caretakers.

By March, even as Tina's and Winkie's pleasurable interactions continued, Tina was developing a special bond with Sissy. Like Winkie, Sissy had been wild-caught at the age of one, in her case in Thailand. Separated from her family and confined to zoos, Sissy experienced a complicated and sad series of events. Swept away in a flood at one Texas zoo, she was beaten by keepers at another. Nonetheless, at the Elephant Sanctuary, Sissy acted in a gentle manner. For emotional security, she carried a tire with her most everywhere she went. But she loved the company of elephants too.

At first Tina made some missteps, pushing, pulling, and poking Sissy in a less than affectionate way. But Sissy's patience was notable, and by April the two elephants were mutually affectionate. During this period, the condition of Tina's feet began to improve markedly. The coincidence of Tina's physical and emotional recoveries makes a lot of sense; as with people, the body and the spirit sometimes heal together. Sanctuary staff thought creatively about how to help Tina, in June even going so far as to take molds of her front feet so that custom-made shoes could be constructed for them. If her tender feet were protected, the staff thought, maybe Tina would begin to explore the richness of the sanctuary grounds. The acres of streams and mud and other mini-habitats belonged to Tina as much as to the other elephants.

These hopes for Tina's future did not come to pass. In July, she died unexpectedly. Under treatment for some minor issues involving loss of motor skills and reduced appetite, Tina had seemed basically fine, and at no time was her situation considered to be life-threatening. She simply collapsed, and she lacked the muscular control to stand even when hoisted to her feet. Lying on a mattress of hay, she stopped breathing.

Tina's human caretakers were in shock and mourned for Tina that day and for a considerable time afterward. It is Tarra's, Winkie's, and Sissy's responses that I want to focus on, though. Tarra was the first elephant to visit Tina's body. Years later, Tarra would become a media star because of her tight bond with a dog named Bella. The "Tarra and Bella" story went viral, spurred by television coverage on *CBS Sunday Morning* and the book *Unlikely Friendships* (see chapter 10). But now, in 2004, Tarra had just lost her elephant friend Tina. So had Winkie and Sissy, and it was those two elephants who stood over Tina's body that entire first night and part of the next day. They refused any chance to leave to take food or water, or a walk. Sissy stood quietly, but Winkie did not; her emotion was apparent in her distraught and repeated prodding of Tina's body.

The next day, sanctuary caretakers gathered to bury Tina. Tarra and Winkie stood at the edge of the grave, where they remained, joined by Sissy, throughout that evening and the next day. Once again, distinct individual differences in mourning were apparent: Tarra was vocal and asked for attention from her human caretakers, Sissy stood vigil, and Winkie paced stiffly around.

On the following day, before moving on to another part of the Elephant Sanctuary, Sissy made a choice that surprised the people who witnessed it. She placed her beloved tire, her security blanket, on her friend's grave. There she left it, an elephant memorial offering, for several days.

6 DO MONKEYS MOURN?

Toque macaques on the island of Sri Lanka live in a visual paradise. The green tree canopy stretches far, and in it the monkeys use their grasping hands to reel in tasty caterpillars suspended from the trees on long, thin threads. The forest boasts lush fruits and a small lake dotted with another favorite monkey delicacy, lily flowers.

Even in the midst of this bounty, the toque monkeys confront dangers, some external to their group and some within it. In a documentary called *Clever Monkeys*, naturalist David Attenborough explains one specific cost of low rank. Only high-status monkeys may hang from tree branches over the water, reaching down to pluck lilies from the surface; their lower-ranking groupmates must enter the water directly and dive for roots and bulbs. The problem isn't only that it takes time and technique to learn how to do this sort of thing, it's also the presence of concrete danger: a large monitor lizard who makes his home in the lake.

Aware of the reptilian danger, the monkeys do post a lakeside guard when low-rankers go into the water. The guard's job is to send up a cry of alarm when the big lizard is sighted. With a vigilant guard on duty, this method works well enough. On the day in question, however, the guard dozes off as a young monkey is foraging in the lake. By the time other monkeys spy the lizard and cry out, it is too late. The camera captures the monitor trundling off with that peculiar side-to-side lizard shuffle, a dead monkey clamped in its jaws. No one follows. The monkey's groupmates attempt no rescue. The lizard isn't mobbed, the monkey isn't visibly mourned.

Later, another toque macaque is shown dead beneath a tree, the loser in a male-male fight for group leadership. His limbs are stiff, his mouth stretched in a mild death grimace. His groupmates approach, including some of his offspring; seven or eight monkeys at once crowd the body. Some lean in and take a sniff, others touch the corpse—when one monkey touches the dead one's locked-upright hand, the hand jerks rigidly back into place. After a while, the curious monkeys move on. The dead leader, beneath the tree, is abandoned.

The monkeys' responses to these two deaths may represent commonplace scenarios among wild animals. The young monkey's death happens swiftly, and the body departs the scene in the predator's jaws. What, if anything, the surviving monkeys think or feel about the event is opaque to us. In the case of the older leader dispatched by his rival, the group response is notable. The body is explored through sight, smell, and touch. To a human observer, it's clear that the monkeys who surround the body know something is anomalous: they surely aren't confusing their dead groupmate with a resting, sleeping, or wounded animal. There are no outward signs of grief.

In the wild, members of tightly knit primate groups experience a great deal of loss. As reported by primatologist Jeanne Altmann in her now-classic book *Baboon Mothers and Infants*, the mortality rate for Kenya's Amboseli baboons approaches 30 percent per annum in the first two years of life. After that, it plunges, but then it rises again, and in adulthood, females suffer a death rate of 12 percent. Though these numbers are specific to certain monkeys during a certain slice of time, demographic profiles suggest they are not unusual for wild animal populations more generally.

Experiencing the death of a groupmate, then, or even of one's offspring or other close kin or social partner, is far from rare for group-living wild animals. If we think about mourning and grief in terms of evolutionary theory, a negative hypothesis (the "null hypothesis," in scientific terms) may come to mind: Wild animals faced with the challenges of survival and reproduction should not expend time or energy on the expression of grief when a group member dies. A weaker version of this same hypothesis would be that wild animals should expend time or energy on grieving only when the resources required for survival are available in sufficient abundance.

If a death provokes no particular emotional response, might this absence of grief be explained as an energy-saving strategy that is under the control of natural selection? If so, might some of the survivors feel emotion but simply ignore it? Or are no emotions felt? By observation alone, without the invasive measures of stress physiology, we cannot distinguish between these alternatives. (We'll consider what those invasive measures do teach us in a moment.)

If any toque macaque is likely to mourn the death of the youngster plucked from the lake by the monitor lizard, it would be his mother. The mother-infant relationship in macaques, as in almost all primates, is exquisitely close. Research shows that in rhesus macaques, a close relative of the toques, mothers and infants share what's called reciprocal face-to-face communication. This suite of behaviors between moms and babies involves smacking of the lips, mouth-mouth contacts, and, most significant of all, sustained mutual gaze.

Think of how important mutual gaze is in our own species, as bonds develop between babies and their caretakers. A memory so vivid that I've carried it for nineteen years comes from my daughter Sarah's infancy. It was a Saturday, exactly four weeks since her birth. I was carrying Sarah in my arms across the street in front of our house, on my way to pay a welcome call on new neighbors. When I glanced down at her, bundled up against the chill November air, she locked eyes with me and let loose with a huge smile. It was what developmental psychologists call a social smile, the kind of aware, intentional smile that is set apart from the reflexive mouth movements of a newborn. To me, a tired but otherwise besotted new mother, the mutual gaze and first social smile meant one thing: my baby was loving me back.

The contours of the emotional relationship between monkey mothers and babies aren't well studied. It's reasonable to expect, though, that gaze and facial expressions shared across the generations both enhance infant survival and cause feelings of comfort or pleasure to flow within the pair. Newborn monkey babies cling to their mothers' bellies; in the beginning, the mother is the infant monkey's universe, the source of all warmth, nutrition, and safety. For the mother, infant care is all-consuming. She starts out carrying the infant on her body around the clock (except in a few monkey species where dads and siblings help out). Moms may bounce their babies, play with them, smack their lips in affection toward them, and try to catch their babies' eye to facilitate that mutual gaze.

That many monkey mothers lose their infants early on is something we know from the mortality profiles. When this happens, some mothers simply put down the body, or leave it where it fell, and carry on with their lives. No visible grief seems to accompany these acts of abandonment. Other mothers, though, continue to carry their infants' dead bodies. Could this carrying be an expression of maternal grief?

Maternal carrying of infant corpses was monitored by primatologist Yukimaru Sugiyama and colleagues for more than two decades in one population of Japanese macaques, close relatives of the toque and rhesus macaques. These monkeys live on the slopes of southern Japan's Mount Takasakiyama. Infant mortality is high, as we would expect in a wild population; based on an intensive nine-year period of data collection, the death rate within a year of rhesus birth was 21.6 percent. Maternal carrying of the dead was observed over a twenty-four-year span, during which 157 cases were recorded out of 6,781 rhesus births. The researchers compiled statistics on factors like infant's age at death and duration of corpse-carrying by the mother. Within a week of death, 91 percent of rhesus the infants had been abandoned by their mothers. The longest maternal carry lasted seventeen days, by which time the small carried body was decomposing, fly-ridden, and reeking with a bad odor. Most of the other monkeys avoided that mother, and when juveniles showed an interest in the decaying body, they were rebuffed by her.

In presenting these data, Sugiyama and his coauthors pose a key question: Does the carrying of dead infants signal maternal emotion or does it instead point to the mothers' lack of awareness that their infants have died? The null hypothesis for this case must take into account the need for wild animals to budget their energy. The carrying behavior does, after all, represent a substantial energy expenditure by the mother. At Takasakiyama, the monkeys must traverse a steep hill every day, and with a dead infant in tow, mothers lose the free use of one hand. Their movement and their foraging are most certainly compromised. So why do they do it? What does it mean that mothers carried infants significantly more often if the death occurred within thirty days of birth? It was especially common if the infant lived more than one day but died within several days; as Sugiyama's group notes, this pattern matches up with the time when the infant, not yet able to move around well on her own, begins to cling and breast-feed regularly. Not all dead infants were carried, however. It's not as if some trigger associated with infant size, weight, or age pushes the mother into an innate response of carrying.

What's most curious to me is that infants who lived longer, and who presumably enjoyed a longer period of emotional connection with their mothers, were not carried longer than infants who were barely known by their mothers. Taking all of the data together, I can't see that the behaviors described for these monkeys fit comfortably with a claim of monkey grief.

Maternal corpse-carrying behavior has also been described by Peter Fashing and his colleagues, who study the gelada monkeys of Guassa, Ethiopia. Large-bodied and long-haired, the Guassa geladas dwell in the grasslands of the Ethiopian highlands. Over a three-and-a-half-year period, fourteen females at Guassa carried dead infants, some for only an hour, others for much longer. Most carrying episodes lasted between one and four days, with three females carrying their infants for significantly longer periods: thirteen, sixteen, and forty-eight days. In these extended cases, the infants' bodies gradually became mummified, and as with the Japanese macaques at Takasakiyama, they emitted an unpleasant smell.

Forty-eight days is a long time to carry a dead body, and suggests to me a decisively willed action on the part of that mother. She resumed reproductive cycling while carrying her dead infant, and was even seen copulating while clutching the infant's body with one hand. The timing of carrying and then abandoning the body cannot, in this case at least, be explained by hormonal changes that occur when an infant suddenly stops nursing. This mother carried her dead infant right through that period and beyond.

Beyond the length of the carrying episodes, what's striking at Guassa is the interest shown in the corpse by females other than the mother. In two cases, juvenile females were allowed to carry and groom the bodies of infants belonging to adult females in the group. Relatively small gelada groups forage independently during the day, but reunite and cluster together on nighttime sleeping cliffs. In one notable instance, Fashing's

HESTER WITH THE DEAD INFANT HISHAM. *PHOTO BY RYAN J. BURKE.*

team saw a female carrying a dead infant belonging to a mother from another group; this female groomed the body and allowed a juvenile female to do the same.

Even keeping in mind the conservation-of-energy hypothesis for wild animals, I found it surprising that mother monkeys in general don't show more discernible evidence of grief. During my fourteen months in Kenya, I found the Amboseli baboons to be closely attuned to each other, smart and strategic in their actions, and ready to defend allies and friends. In reading published accounts and talking to other primatologists, though, I was forced to conclude that little evidence for monkey grief has emerged from observation alone.

In fact, the report by Fashing and his colleagues contains descriptive passages that increase my sense of caution about concluding that monkeys mourn. Two geladas, a mother-infant pair named Tesla and Tussock, died in April 2010. Tesla, the mother, had been severely weakened by sickness that followed from a parasitic infection. During the period of her illness, two younger females helped her out by carrying Tussock, her seven-month-old daughter. But when Tesla became too ill to leave the group's sleeping cliff, the other geladas departed to forage without her. Slowly, Tesla and Tussock managed to move to a spot about 175 meters from their sleeping position. When the group returned to the cliff that night, Tesla and Tussock's new position wouldn't have been visible to them. None of the geladas showed any apparent concern for their missing groupmates, and none searched for Tesla and Tussock. The next morning, the primatologists found Tesla dead. Throughout that day, Tussock, now on her own, stayed by her mother's body, "crying plaintively and rocking side-to-side." The next morning, the infant too was found dead.

It seems likely to me that Tussock felt something upon her mother's death. How could she not have felt afraid, left alone in the cold, outside the protective web of her group, with her mother lying inert and unresponsive? If she felt grief, she suffered it alone. When I asked Tyler Barry, one of the primatologists at Guassa at the time of these events, for his interpretation of what Tussock might have been feeling, he said, "I would not feel comfortable arguing that Tussock was feeling grief as she cried and rocked. I am pretty sure she hadn't had milk for almost two days at that point and was probably dehydrated and reaching starvation point. I am sure the cold is what killed her in the end, though, and also may have been a cause of the rocking."

Barry confirmed that Tesla and Tussock's location, away from their sleeping cliff, meant that their suffering would have gone unnoticed by their group. "There was a bachelor group that glanced down at Tesla's dead body the morning of the second day," he recalled, "but other than that they were too far away from the normal sleeping spot for the main herd to even hear Tussock." Gelada males, then, who were not Tesla's regular associates, showed a brief curiosity response to her body, but the monkeys who might have mourned were too far away.

If grief does accompany the act of maternal corpse-carrying in monkeys, or an infant monkey's solitary vigil at the side of her dead mother, we cannot know it from observation. So far, the null hypothesis, which predicts no great expenditure of energy on grief by wild monkeys, remains formidable.

Primatologists Dorothy L. Cheney and Robert M. Seyfarth, among the world's leading experts on wild monkey behavior, take note of monkeys' lack of visible grief in their book *Baboon Metaphysics*. When monkeys carry infants who are dying, these scientists say, they treat them pretty much the same way as they do their healthy infants. Cheney and Seyfarth embed this observation in a broader context. Monkeys do not share food with sick companions, nor do they come to the aid of elderly or disabled group members. Baboon mothers specifically, they write, "often show a surprising lack of concern for their offspring's anxiety and distress during water crossings or at other times of separation."

When mothers carry not a dying but a dead infant, other baboons show interest, as we've seen with other monkeys—but an interest of a limited nature. "In the minds of other group members," Cheney and Seyfarth write, "the infant's status seems to change soon after it dies: they cease to treat it as an infant." The baboons inspect the corpse but never direct grunting vocalizations to it, as they would were the infant alive. They do not try to pry the corpse from the mother. Most interestingly, if the mother puts down the corpse and moves away, a close relative or male friend may guard the baby until she returns. If scientists approach the corpse with the aim of obtaining a DNA sample, group members may threaten them. Cheney and Seyfarth conclude that the baboons' response is not an expression of grief or empathy but instead is organized around ownership—the idea that the infant once belonged, and indeed still belongs, to a particular female and to the social group as a whole.

What happens when we add physiological measures to straightforward observation? Cheney and Seyfarth carried out their long-term baboon research at the Moremi Game Reserve of Botswana's Okavango Delta; one study done under their supervision adds some biochemical insight to the question of monkey grief. Like the baboons at Amboseli, the Okavango baboons live in multimale, multifemale troops, and female relatives organize themselves into close-knit groups called matrilines. Grandmothers, mothers, daughters, aunts, nieces, and young sons and nephews all spend time close together in social grooming and social alliances. At puberty, the males transfer into another group. This pattern means that in any given group, the adult males tend to be strangers to each other, unlike the related adult females.

In the Okavanga group, as at Amboseli, predation is quite high. During a sixteen-month period in 2003 and 2004, twenty-six baboon deaths were recorded. All but three were healthy animals. Ten were known to have been taken by predators, based on researchers' direct observation of the attack or by the presence of the body; the remaining thirteen were suspected to have met the same fate, based on predator sightings or alarm calls vocalized by the monkeys.

Living amid this sort of danger, the Okavango baboons become stressed, and that stress shows up in their bodies. Researcher Anne L. Engh and her coworkers collected fecal material from the female baboons in order to measure levels of the glucocorticoid (GC) hormone, a type of stress hormone that circulates in the body, then is excreted through bodily waste. The researchers found that in the four weeks following a predation event in the group, females' GC levels increased measurably. This finding makes intuitive sense; imagine witnessing a lion or a leopard stalking your circle of family and friends, then singling one out and and making a kill. Stress hormones in our bodies would flare up too under conditions like these.

Probing further, the research team discovered the chemical signature of grief in the baboons. The GC levels of twenty-two "affected females," each of whom had lost a close relative to predation, were compared to those of a control group made up of females who had experienced no such loss. The affected females showed significantly higher GC levels. Engh and her colleagues emphasize that while predator attacks were witnessed by many adult females in the group, only the "bereaved" females showed significantly higher levels of GC.

The bereaved monkeys' elevated stress levels lasted only four weeks, perhaps because the females soon began to increase the number of their grooming partners and the rate at which they participated in grooming. In monkeys, to groom and be groomed by a partner is a soothing social activity as much as it is a hygienic one. As Engh's team put it, "Bereaved females attempted to cope with their loss by extending their social network." While it's rash to make facile comparisons between monkeys and humans, I can't help but think of the person who, after mourning a loved one, gradually reaches out to potential new friends in the community, church, or workplace.

I was also struck by Engh and her coauthors' willingness to use the word "bereaved" in their scientific publication. Here was the first clear indicator that I had come across of monkey grief, and it was approached in terms of physiology rather than of social behaviors that might indicate mourning. But when I asked Engh whether any of the Okavango female baboons had shown signs of grief, I learned something that had not been mentioned in the peer-reviewed article. Engh shared her recollection of Sylvia and her adult daughter Sierra, two baboons who were unusually close. "They groomed each other almost exclusively," Engh told me, "and spent much of their time together." Then Sierra was killed by a lion. To Engh, Sylvia appeared depressed. She sat apart from the other baboons and did not initiate social interactions. This went on for a week or two. "Sylvia was high-ranking and intimidating," Engh says, "so it didn't seem unusual that other females didn't approach her, but I was surprised that she had no apparent interest in interacting with anyone." In fact, it was Sylvia's behavior that led Engh to initiate the GC study. Sylvia had been close with her daughter, and when that closeness was stolen by death, it brought her to grief. As fits the monkeys' pattern, Sylvia's altered behavior lasted only a few weeks. She then widened her social circle by befriending other females.

An obvious place to look for mourning is among mammals who, unlike macaques, geladas, or Okavango baboons, organize themselves into pair-bonded couples. Among birds, pair-bonding is a routine occurrence, but only 5 percent of mammals do it. One exception is a rodent called the prairie vole, and scientific work on the biological and emotional basis of pair-bonding in these small animals may help us think about grief in monkeys.

In experimental work published in 2008, Oliver J. Bosch and his colleagues decided to look into how even short separations from mated partners would affect prairie vole males. The males to be studied were paired either with females they had never met before or with male siblings they had not encountered since the time of their weaning (a period that ranged from forty-nine to seventy-nine days). After five days together, half of these pairs were separated.

All of the study males were then subjected to stress tests, including what's called a forced-swim test, in which the vole remains for five minutes in a beaker of water; a tail-suspension test, in which the vole is hung for five minutes from a stick to which its tail is taped; and an elevated-maze test, which tests the vole's inherent fear of exposed spaces, again for five minutes. In two of these tests, males who had been separated from female partners showed increased levels of "passive stress coping," a type of response that correlates to higher depression levels. In the forced-swim test, they tended to float rather than struggle or swim, and in the tail-suspension test, they hung passively. The effect was specific to males separated from females, as opposed to either males paired with male sibs or males kept isolated. It's the specificity of the effect that's important here: it shows that it's the pair bond that matters most, emotionally, to these voles.

Further, these scientists found that the stress responses were mediated by what's called the corticotropin-releasing factor (CRF) system. Levels of CRF, thought to mediate anxiety and depression, rose in the males separated from female partners. That ameliorating effect might sound like a good thing for the stressed males, but could the anxiety and depression in fact be adaptive in some way? Bosch's coauthor Larry Young explained to me that when CRF receptors were experimentally blocked, the voles didn't show the depressive behaviors. And yes, Young believes that the entire system is adaptive for the voles: "The negative state produced from separation from the partner," he told me, "serves to drive the male back to their partner, maintaining the pair bond."

I experienced an emotional response of my own in reading about these experiments. Five minutes, the length of time a male was forced to swim or hang by his tail, isn't an eternity, but still I began to wish I'd served on the animal-care committee that approved these tests. Then I read further. In pursuit of answers to the CRF-receptor questions, a number of voles had been decapitated. While these experiments violated no institutional ethics policies, they gave me great pause in weighing the costs to animals of our invasive probing of their emotion (versus behavioral observation or analysis of, say, fecal material). Bosch and his colleagues believe their vole work, both the behavioral and biochemical components, may illuminate the expression of human bereavement. This hope may be realized. But perhaps the voles themselves, when a partner dies, experience grief and bereavement. It doesn't seem that this question was asked.

Let's return to monkeys. None of our closest living relatives, the great apes—chimpanzees, bonobos, gorillas, and orangutans—pair off to raise their young. As noted earlier, this is par for the course among mammals. But the so-called lesser apes, the gibbons and siamangs, do bond in this way, as do some monkeys, including titis, owl monkeys, marmosets and tamarins.

Even in the context of monkey monogamy, emotion hasn't been well studied. In South American titi monkeys, males and females form bonds for life. When a male and female pair is forcibly separated by scientists in the laboratory, the animals show their distress through agitated behavior, and their plasma cortisol levels rise. In a comparative study by Sally Mendoza and William Mason, squirrel monkeys subjected in the same way to male-female separation showed no comparable behavioral or physiological changes, presumably because, unlike titis, they are not monogamous. The pair bond in titi monkeys, in other words, isn't merely a matter of survival and reproductive success—the monkeys matter to each other.

The same question comes to mind with monkeys as with the voles: How might we move from lab findings that rely heavily on blood chemistry and basic measures of behavioral distress to understanding more about what survivors of a disrupted pair bond experience? A video archive could help answer this question, with the gold standard defined as filmed records shot around the time of death of one monkey, focusing on the surviving partner and other family members. In this way, scientists could make fine-grained behavioral comparisons among monogamous and nonmonogamous, and captive and wild, monkeys.

Rare events in the lives of monogamous monkeys might be hard to film in the wild. These species are largely arboreal, and as researcher Karen Bales remarked to me, with tree travel it may simply be too difficult for mothers to carry a dead infant or for a monkey to stay in proximity to a dead group member. In captivity, filming such events would be easier. I predict that grief responses will be found to occur among some pair-bond survivors in captivity, but testing of this hypothesis is sorely needed.

At Miami's DuMond Conservancy, owl monkeys Betsy and Peanut lived together as pair-bonded partners for eighteen years. Peanut, who was born in the wilds of Peru, was originally shipped to a US research laboratory as a test subject, but after some years and a serious illness he was allowed to retire to the conservancy. At first, he was timid around others of his kind. Then he met Betsy. Primatologist Sian Evans notes that even for owl monkeys, "Peanut and Betsy were unusually closely bonded. They had several offspring, and Peanut was a devoted father, carrying and caring for them all."

Then, in 2012, Peanut began to weaken. He still foraged for insects at night, right alongside Betsy (owl monkeys are the only monkeys who are nocturnal). But the quality of his movements indicated his frail state, and eventually he became ill. When attempts to treat him failed, the staff returned him to his enclosure, where he could live out his remaining hours with Betsy. As always, Betsy was an attentive partner. She held and nuzzled Peanut until he died.

Without Peanut, Betsy immediately began to behave differently. For the first time, she sought out and spontaneously interacted in a friendly way with Evans, her caretaker. In the past, Evans told me, Betsy had viewed her as a competitor, as female owl monkeys will do with female humans. But with Peanut gone, there was a behavioral shift. Evans grieved for Peanut and found comfort in Betsy's companionship. What Betsy felt, Evans says she cannot know, and she doesn't feel comfortable calling Betsy's response "grief." She refers to it instead as "response to loss," a sort of interspecies shift in bonding that came about as a result of Betsy's being without her partner.

So then, do monkeys mourn? Often they do not, at least in a way visible to us. This conclusion holds even in some cases where, from an anthropomorphic perspective, it seems that they should mourn—as with mothers who carry their infants' corpses for weeks, or the owl monkey Betsy who lost her pair-bonded partner of eighteen years. But some monkeys do grieve, as Anne Engh's description of Sylvia the baboon tells us. A more robust conclusion awaits the day when statistical or physiological profiles are complemented by vivid descriptions of monkeys' behavioral responses to death.

7 CHIMPANZEES

CRUEL TO BE KIND

As the world's first "astrochimp," Ham the chimpanzee rocketed into outer space in 1961. Captured from Cameroon, West Africa, brought to the United States, and named for the Holloman Aerospace Medical Center, Ham flew 155 miles above the earth's surface at five thousand miles per hour on behalf of the American space program. Inside the Mercury capsule, Ham carried out the tasks for which he had been trained; as lights flashed, he pulled levers in response, demonstrating that the effects of space travel do not harm a thinking primate's capacities. Ham thus cleared the way to send humans into orbit.

Few ethical concerns were raised back then, more than a half century ago, about subjecting an ape to this sort of stress. In hindsight this nonchalance seems cavalier, especially given the safety record of previous primate launches. A monkey called Albert I suffocated during his flight in 1948. The following year, Albert II died from impact trauma when his capsule's parachute failed on returning to earth; Albert III was killed when his rocket exploded at thirty-five thousand feet; and Albert IV was killed on impact in another parachute failure.

There's something chilling about this litany of cloned names, a sense that monkey after monkey was sent to his death with no thought to the individual lives risked and lost. Indeed, a name was not bestowed on Ham the chimpanzee until he returned safely to Earth, for fear that his human coworkers in the space program would get too attached to him. In subsequent years, following the quartet of Albert deaths, the record of monkey survival did improve. As late as 1958, though, a monkey died because his capsule could not be located following splashdown in the Atlantic.

The BBC's coverage of Ham's successful flight on January 31, 1961, reflects the lighthearted tone of the day, even as the reporter describes the unexpected outcomes of the Mercury mission:

> Because of his steeper-than-expected climb, the capsule overshot its landing site in the Atlantic off Florida by some way. Ham had an uncomfortable three-hour wait before he was found. Then when rescue helicopters finally arrived, they found the capsule on its side and sinking. It had landed with such force that the heat shield had punched two holes in the capsule. Ham, however, took it all in his stride and when the spacecraft was opened accepted an apple and half an orange in reward.

Jane Goodall had, by 1961, already begun observing wild chimpanzees in Tanzania. But the world didn't yet know about chimpanzees' deep family bonds, their clever making and using of tools, their emotional capacities for love and grief. Looking back, given what is known now, we can only wonder: did Ham really take it all in his stride? Or was he terrified, both because of intense heat and because he was bobbing around untended in the ocean for three hours, not knowing what would happen next? The image is hard for the mind to take, Ham alone in the capsule, with no other being to empathize or to comfort him during what can only have been a truly frightening experience.

Many years later, at the National Zoological Park in Washington, DC, Ham responded to Melanie Bond, a biologist and ape-keeper, in a way that spoke to her of empathy and comfort. Ham had been retired from the space program into zoo life; for a long seventeen years, he lived in the nation's capital as the zoo's sole chimpanzee resident. (His final years, fortunately, were spent in apparent contentment with other chimpanzees at the North Carolina Zoo.) Bond, over the next decades, would spend countless hours caring for great apes both there and at the Center for Great Apes, a sanctuary in Florida, fostering a particularly deep affinity with orangutans. At the time of the incident with Ham, though, in 1977, she was relatively new to the zoo.

The first deep connection Melanie developed with an ape had been with Archie the orangutan. One day, it was her job to assist in his routine physical, for which he was given an immobilizing drug in his cage. During the examination, Archie stopped breathing. The zoo's veterinarian, Mitchell Bush, made a heroic effort to bring him back. For forty-five minutes, Dr. Bush compressed Archie's chest, offering CPR with such vigor that Archie's sternum cracked. Eventually, all present at Archie's side were forced to accept that the orangutan had died.

As Melanie walked away, past a line of caged apes who could see her clearly, she cried for Archie. She remembers it as quiet crying: nothing terribly overt, just tears she couldn't stop. Melanie saw Ham watching her and said aloud to him, "Yes, Ham, I am very sad." Moving slowly and gently, Ham reached a thick finger through the cage bars and touched a single tear on Melanie's cheek. He then smelled and tasted the tear. "I felt empathy," Melanie remembers. "I felt, 'someone understands.'"

What would the skeptics say? That Melanie projected her need for comfort onto a chimpanzee's actions? Sure, they might admit, Ham was a smart ape, and curious about Melanie's crying. It was this curiosity that motivated his actions, not any emotional resonance with Melanie's state of mind, nor any desire to give solace to a friend. To impute such resonance and desire to a chimpanzee is to engage in wishful thinking about humans' and apes' emotional similarities.

In support of the skeptic's point, films of chimpanzees in the wild rarely include scenes that mirror the gentleness exhibited by Ham. The iconic still photograph of a wild chimpanzee, taken at Gombe, Jane Goodall's research site in Tanzania, shows an ape who has fashioned a wand tool inserting it into a termite mound to retrieve a protein snack. The videocamera, though, gravitates toward outbreaks of aggression, displays of hotheaded chimpanzee excitement. These recordings lead to descriptions of brutality, as in the case of an attack filmed and narrated by the anthropologist David Watts of Yale University. Watts observed the event among the Ngogo community of chimpanzees of Kibale National Park, Uganda. In the key part of Watts's footage, a pack of chimpanzee males is seen to surround, then begin to kick and bite, another male called Grapelli. Hurling themselves at the cowering figure, the males inflict injuries so severe that Grapelli died three days later.

The action of the male chimpanzee mob may seem shocking to us, and may tempt us to label the entire species as violent. After all, similar attacks have been seen in other chimpanzee populations, and they differ appreciably from the hunting behaviors that occur when, for instance, chimpanzees catch and consume colobus monkeys. In Kenya, when I'd return to my house after following baboons all day through the bush, I would often hear the roars of lions through the open-mesh window of my bedroom. I'd be haunted by the thought of zebra, antelope, and wildebeest, out there on the savannah at the big cats' mercy. Lions ate "my" baboons, too, but with a barking alarm upon sighting a tawny stalking shape, the monkeys would scatter high up into the trees, with at least some hope of escape. Not every monkey always made it, but arboreal refuge did exist for them. Not so for the herbivores, who could run but never hide: for them, no tree, no burrow, no watery hiding place.

Even so, when a lion brings down a zebra, or a fox snatches a rabbit, the animal hunters aren't stigmatized as violent. But what of these Ngogo chimpanzees filmed by Watts? They are shown engineering, with shrieking abandon, the horrific end of one of their own; Grapelli belonged not just to the same species but to the same community as the attackers.

Watts does describe the empathetic response of one of the males who refused to join in the attack mob, and who stayed near Grapelli as much as he could. The vast majority of males, though, showed no such mercy, and certainly nothing like the gentleness shown by Ham to his human friend at the National Zoo. Is it just that in Ham's case, the chimpanzee drained out of him after he was stolen from his Cameroonian home? So long subjected to the press of human needs—first as a guinea pig in the space program, later as entertainment for zoo visitors—did Ham become only the palest reflection of a wild chimpanzee?

It's easy to be swept away by the picture of chimpanzee aggression I've painted here, but there's another side to wild chimpanzees too, a side that brings us much closer to Ham. The expression of chimpanzee grief, in the wild as well as in captivity, has complicated the picture of what is understood to be "natural" for chimpanzees. Arguably the most famous example of grief in the animal world, dated to 1972, is the juvenile chimpanzee Flint's loss of will to live following the death of his mother Flo.

At Gombe, Flint had enjoyed his mother's undivided attention well beyond his infancy. His younger sibling Flame—Flo's last-born offspring—had died, leaving him the emotional center of his mother's aging life. As Goodall wrote in *In the Shadow of Man*, except for nursing (because Flo's milk had dried up), "Flint again became Flo's baby. She shared her food with him; she permitted him to climb onto her back or even on occasions cling to her belly. She groomed him constantly and, as of old, she welcomed him into her bed at night." These patterns of behavior persisted until Flint was over six years old, and even after that point, mother and son remained abnormally close. Flo died when Flint was eight years old, and he was unprepared to cope. In a passage from Goodall's later book *Through a Window*, we glimpse the depth of Flint's loss: "The last time I saw him alive, he was hollow-eyed, gaunt and utterly depressed, huddled in the vegetation close to where Flo had died. . . . The last short journey he made, pausing to rest every few feet, was to the very place where Flo's body had lain." Only three weeks after his mother's death, Flint died too, of causes that Goodall unflinchingly attributes to depression and the resultant weakening of his immune system.

When the situation is reversed—when, against the natural pattern, mother chimpanzees lose their loved offspring to death—that loss too may be felt acutely felt. Just as wild monkey mothers do, chimpanzee mothers sometimes carry their babies after death. And occasionally, just as with monkey mothers, a chimpanzee mother may seem unable to stop the corpse-carrying, even though the body is rotting in her hands.

In chimpanzees, emotional ties between mothers and their offspring may be intense. In the wild, great ape babies nurse from and ride on the mother for four years or more. When these babies die, they may continue to ride on their mother's body, simply because the mother refuses to part with them. At Bossou, in the West African country of Guinea, a respiratory epidemic swept through the chimpanzee community in 2003. Two infants under the age of three, Jimato and Vene, were among the victims. Their mothers, Jire and Vuavua, carried the infants' bodies for sixty-eight days and nineteen days respectively. I find Jire's sixty-eight-day commitment to her infant's dead body astonishing; just think of her burdened with that tiny corpse for an entire summer, the equivalent of July Fourth to Labor Day and a little bit beyond.

Vuavua's nineteen-day carry, as it turns out, overlapped with Jire's

WILD CHIMPANZEES IN UGANDA: ALPHA MALE NICK, FEMALE KALEMA, AND KALEMA'S FIVE-YEAR-OLD SON. *PHOTO BY LIRAN SAMUNI.*

longer one. Did one mother pass the other in the forest, look into her eyes, and acknowledge a shared loss? Did each mentally revisit times when her child was alive, nursing close against her body? Intruding into these sentimental thoughts is a grim reality. The mothers must have experienced terrible sights and smells, judging from a report by primatologist Dora Biro and her colleagues. The infants' bodies became mummified, just as the monkey infants' corpses that we discussed in the previous chapter had: the hair was lost, and the limbs and other body parts turned leathery. "Because of wearing effects of prolonged carrying," Biro's team wrote, "by the time Jire abandoned Jimato's body, much of the body cranial structure had been destroyed, making most facial features unrecognizable."

The mothers shooed away flies from their babies' bodies, and even groomed the corpses. Sometimes, infant and juvenile chimpanzees were allowed to borrow a body and carry it in playful ways. Do such actions mean that the mothers persisted in their carrying because they couldn't discern that the infants had died? I doubt it. For one thing, the carrying techniques used by the moms varied greatly from those normally used with healthy infants. For another, chimpanzees are capable of complex reasoning, thinking strategically, step by step, about how to solve foraging problems with tools, or social challenges with deft manipulation of allies. While it's impossible to prove that chimpanzees understand anything about death, it's equally impossible for me to think that chimpanzee mothers could judge the dead babies—unbreathing, unfeeling, and rotting—to be alive.

Of course, Jire and Vuavua are females. The brutality that I described earlier, visited on Grapelli in the Ngogo community, was orchestrated by males. Beyond the behavior of juveniles like Flint, is there a place for male sensitivity to death among chimpanzees living in the wild, a gentleness akin to Ham's expressed in captivity?

In the first-published scientific review of primate death, James R. Anderson describes an event that took place in Cote d'Ivoire, West Africa, in 1989:

> In the Tai Forest, a fatal leopard attack on an adolescent female chimpanzee caused an outburst of loud calling and aggressive displays by males, who initially dragged the body over short distances. . . . Contacts with the body were frequent, including grooming and some gentle shaking. Interestingly, infants were prevented from approaching the body. Again, after several hours the corpse was abandoned.

In a bare-bones sense, Anderson's summary is accurate. Yet it misses the nuance—indeed, the significance—of what happened that day at Tai. Anderson's selective reporting echoes a situation that I noted in chapter 6. There I cited a paper by Anne Engh and colleagues that reported a hormonal spike in bereaved female baboons relative to other baboons. Devoted to statistical results, the paper included no observations of grief in the monkeys. Its genesis, however, was Engh's witnessing of grief symptoms in a female whose daughter had been killed by a predator. The peer-reviewed scientific literature—including Engh's and Anderson's papers—favors statistical results and bare summaries over descriptive passages. Yet it is in descriptive details that a topography of animal grief will emerge.

Precisely this needed detail can be found in *The Chimpanzees of Tai Forest* by primatologists Christophe Boesch and Hedwige Boesch-Achermann.

There the anecdote summarized by Anderson is presented in full, and the narrative strongly suggests that even adult male chimpanzees may respond to a companion's death with thought and compassion. In the forest, the chimpanzee Tina was found dead by Tai field assistant Gregoire Nohon. Viscera protruded through her stomach. An autopsy later showed that she had died when a leopard bit through her the second cervical (neck) vertebra. Four months previous, Tina's mother had also died. Since that time, Tina, age ten, and her little brother Tarzan, age five, had been traveling with Brutus, the community's alpha male. From what they could see of Tarzan's actions, Boesch and Boesch-Achermann concluded that he longed to be adopted by Brutus. Sometimes, he even shared Brutus's night nest. Only after the discovery of Tina's body did these two scientists realize just how tight the ties were within the trio of Tina, Tarzan, and Brutus.

The Boesches found a dozen chimpanzees, six females and six males, sitting in silence around the body. Over the next hours, some of the highly aroused males performed displays around the corpse. Some touched Tina. In a period of eighty minutes, males Ulysse, Macho, and Brutus groomed the body for nearly an hour. Ulysse and Macho hadn't been seen to groom Tina at all when she was living; other males in the community had groomed her only for brief periods. Here, then, was a new and unexpected behavior. Further, some chimpanzees gently shook Tina's body, as if trying to solve the puzzle of her stillness.

Other chimpanzees played gently together near the body. To engage in play around a dead companion may seem a strange choice, but how many of us have joined in joking and laughing during the long hours of a wake or memorial service? The urge to play verbally may be a natural way to remember happy times spent with the one now deceased, or it may represent a discharge of nervous energy at an emotionally trying time. The Boesches think that perhaps the chimpanzees needed to expel the tension caused by Tina's violent death, and that playing and even laughing near the corpse allowed them to do this.

About two and a half hours after Tina's body was first discovered, Tarzan walked up to his big sister. By this point, other young chimpanzees had been run off by Brutus, who acted as a sort of gatekeeper. "Tarzan came to smell gently over different parts of the body," Boesch and Boesch-Achermann reported, "and he inspected her genitals. He was the only infant allowed to do this." Tarzan groomed his sister, and pulled on her hand. While this scene was playing out, Brutus chased Xeres and Xindra, a mother-daughter pair, from the area.

That Tarzan spent time with his sister, unlike other young chimpanzees and even some adults, was no random outcome. Brutus brought it about in a thoughtful way. An eminently smart chimpanzee, Brutus played a key role in the Tai community, particularly as a hunter. At Tai, the males hunt cooperatively for monkey. The "moves" needed for a successful capture in the thick forest take years to learn, especially because multiple males must work together strategically, taking specific, conscious steps to aid each other rather than charging in "every male for himself" and hoping for a good result. Mastering hunting skills among the Tai chimpanzees takes twenty years, with the most complex skills taking another ten years more.

And Brutus was a star hunter. He was, during the observation period highlighted by the Boesches in their book, the best meat provider of the community. His cognitive feats were unmatched by any other male, especially his performance of what is called double-anticipation hunting. On a number of occasions, Brutus mentally calculated not only the imminent moves of his monkey prey, but also those of his fellow chimpanzee hunters. In anticipating their actions, Brutus showed that he could reflect upon the mental state of others. In scientific lingo, Brutus has a theory of mind; he bases his own actions in part on an awareness that other intelligent creatures may act or feel in ways different from his own.

These capabilities were, I believe, fully engaged on the day of Tina's death. Brutus recognized that Tarzan, alone of all the young chimpanzees at Tai, needed time to inspect his sister's body and to mourn over it. Unlike Flint, grieving alone over his mother Flo's body, Tarzan mourned as part of a social community, because the alpha male of that community recognized his relationship with his sister. I would even venture to call the Tai chimpanzees' response to Tina's death a "wake" of sorts, because so many apes gathered around the body.

All told, chimpanzees stayed constantly with Tina's body for six hours and fifteen minutes. Brutus himself spent four hours and fifty minutes with her, only interrupting that vigil for seven minutes. Eventually, the chimpanzees left the area around Tina's body. Two days later, a leopard consumed part of the carcass. In this way, becoming part of another animal, Tina vanished back into the natural world. Or to put it another way, she remained part of the natural world. We are left to wonder what Tarzan, Brutus, and her other social partners thought about or remembered of Tina in the weeks and months that followed.

In *Ape*, a volume in the fabulous "Animal" series published by Reaktion Books, John Sorenson notes the strange blend of factors that makes up humans' response to our closest living relatives: "Although much effort goes into denying our proximity to other apes," Sorenson writes, "we are fascinated by their resemblance to us and by possibilities of transgressing the border separating us." We stare at chimpanzees, bonobos, gorillas, and orangutans in zoos and on film, seeing almost but not quite human selves looking back at us. Watching movies, TV shows, or commercials, we may chuckle at chimpanzees dressed in clothes, drinking out of teacups, or wearing suits and carrying out tasks in a modern office. Often in these scenarios, something goes slightly amiss, and the rules of everyday life are broken. "Watching the chaos," Sorenson tells us, "we can see, in a safely managed way, what things would be like if we did not maintain control of the situation and of ourselves."

And that's the point—we don't always maintain control of the situation or ourselves. In one sense, times are changing: fewer of us laugh at chimpanzees out of control on the screen, while more of us protest the entertainment industry's unethical treatment of the apes. But that laudable sea change does not alter the reasons for the enduring popularity of the chaotic ape skit. As Sorenson hints, it's that sense of being on the edge of chaos, of giving in to our own wild impulses, that may explain our fascination. Primatologist Frans de Waal has described our species' Janus-faced tendencies, our equal capacity for compassion and for cruelty. Our nature, he says, is split between two ancestral sides; we share a common ancestor with both excitable, violent chimpanzees and calmer bonobos, who are regarded as peacemakers. I could just as well, though, shift the frame and focus on individual variation from chimpanzee to chimpanzee. Like captive Ham in his empathetic response to Melanie Bond's grief, we humans shine with goodness; like the wild chimpanzees filmed by David Watts, we explode with violence, causing pain and sorrow to others, sometimes on the scale of genocide.

I do not believe that people act in these conflicting ways because the patterns are inherited, a fixed part of our nature. Anthropologists' work with people around the world, and back through time, convincingly demonstrates that there is no single human nature. Our evolutionary legacy is to behave, think, and feel flexibly, according to what happens around us in combination with influence from our genes. We construct our natures in response to a web of experiences that spans from cradle to grave. In a similar though less elaborated way, apes' variable behavior, and its responsiveness to life experience, tells us that there is no single chimpanzee nature (or bonobo nature, or gorilla or orangutan nature).

Some chimpanzees kill fellow community members in brutal ways. Some chimpanzees mourn others in their group and express compassion for others who mourn. I wouldn't be surprised to learn that the chimpanzee who participates in a violent male mob is capable also of mourning. Chimpanzee grief is real, just as chimpanzee violence is real.

8 BIRD LOVE

Every March, a stork flies from South Africa to a small village in Croatia, a distance of eight thousand miles. This bird, who has been given the name Rodan, times his arrival with astonishing consistency, alighting each year in the village on the same day and at about the same hour. In 2010, his fifth year making the journey, he showed up two hours earlier than usual, surprising the small crowd of people gathering to await his return.

But it's not people that Rodan flies so far to see. It's his mate Malena, a stork who, years earlier, had been shot by a hunter. Malena's injuries prevent her from joining Rodan on his annual migration. A kind man in the village cares for her and reports that each year the two birds visibly delight in their reunion. Rodan and Malena make good on their affection, too; at least thirty-two chicks have been born to the couple. It's Rodan, of course, who tutors the young ones in how to fly. And when the pull of the southern hemisphere takes over, the fledglings accompany him to South Africa.

Video footage shows the two companions grooming and mating on a rooftop in the village. How must Malena feel, left behind year after year by her mate and her offspring? Does she remember, and miss, soaring sky-high across the globe? And what are the reasons for Rodan's persistent loyalty, his preference for Malena over all other storks? Has he imprinted upon Malena in some adult equivalent to the attachment behavior famously shown by baby geese for Nobel Prize–winning animal behaviorist Konrad Lorenz? Or do we have here an example of bird love, which will lead inevitably to grief for the survivor when one of the pair dies?

Bird bonding can be a funny thing. Sometimes it goes awry. Petra, the sole black swan to live a lake in Munster, Germany, has bonded not to another swan but to a swan-shaped white plastic pedal boat. The boat is so integral to Petra's emotional well-being that when she was sent to a zoo, the boat went with her. In telling Petra's story in his book *The Nesting Season*, Bernd Heinrich shows there's a lot of instinct involved in bird attachment. According to Heinrich, storks like Rodan and Malena tend to bond to the nest site more than to each other. Based on stork biology, then, if Rodan arrived in Croatia some year to find a strange female stork in the rooftop nest, he would probably groom her and mate with her, and faithfully return to her next year.

Is one opposite-sex stork, then, as good as any other? Is the newspaper writer who called this male "Rodan the Romantic" going too far? We are willingly beguiled by such stories, and it's not just storks. When the now-famed nature documentary *March of the Penguins* came out, people flocked to the theater in droves to revel in scenes of warm and fuzzy bird coparenting. Why are we collectively attracted to stories of loyalty on the wing, of bird love, and hope for them to be about genuine emotion more than mere instinct? Could this intense interest in bird bonding relate to our species' fraught relationship with monogamy?

Now, my best and first reader is my husband, so let me clarify that, twenty-three years into our marriage, I'm as happy to see Charlie after we spend a few days apart as Malena is to see Rodan after his overseas flight. That's just the way it turned out for us. Mounting evidence suggests, though, that monogamy has never been a natural state for our species. There's no evidence that nuclear families, centered on a male-female pair bond, were part of our evolutionary past, and even in modern societies they are a minority pattern. To stay with one partner exclusively over the long haul is relatively rare for *Homo sapiens*; why it remains a cultural and emotional ideal for so many of us is an intriguing question.

Do we see in faithful bird pairs an ideal, a hope for our own relationships? "In the movie *Heartburn*," biologist David Barash writes, "a barely fictionalized account by Nora Ephron of her marriage to Carl Bernstein, the lead character complains to her father, who responds 'You want monogamy? Marry a swan!'" As it turns out, though, bird pair-bonding isn't as idyllic as we wish it to be. With evident glee at the opportunity to smite a myth, Barash explains that, in fact, swans aren't monogamous and neither are many other birds. In one study, female blackbirds paired with vasectomized males continued to lay fertile eggs. DNA studies show that a lot of supposed "monogamy" actually involves what many of us call cheating but scientists call EPCs, extra-pair copulations. These ornithological data are robust and apply across many species. Is it just so much foolishness, then, to get misty-eyed over Rodan and Malena?

Questions arise, too, when we think ahead to what's in store for the pair. Inevitably, one year, Rodan will not show up at Malena's nest, because he has flown elsewhere, grown too old to make the journey, or died. Or maybe he will show up only to find Malena gone—or guarded by another male stork. Will the bird who is spurned, or left alone, mourn the loss, or just move on to another partner?

When it comes to monogamy, here's the thing: myth-busters like Barash merely sow the seeds of a different myth, the myth that it's naïve and a little bit silly to imagine that birds might care deeply for their partners. But why couldn't long-term bird mates feel something for each other? Malena's caretaker sees emotion when the two birds reunite, and even if some EPCs occur along the way, surely cheating doesn't preclude affection for the original partner. Indeed, scientists make a distinction between social monogamy and sexual monogamy. Animals who don't show sexual fidelity but still stay together as pairs are labeled socially monogamous. This technical distinction makes some sense, but it strips the birds of an emotional life. Compare this scheme to how we think about the adulterous affair, the human equivalent of EPCs in birds.

Penelope, that Homeric ideal of faithfulness, never cheated on Odysseus during his two-decade absence from home. She pined with loneliness, but stayed true in her body and her heart. Odysseus himself achieved no such spousal fidelity—remember the temptress Circe? We might playfully suggest that Homer, in creating a long-loyal woman and a philandering man, anticipated modern-day pop-psychology stereotypes (the man strays, the woman stays). But despite his adultery, no one thinks to question Odysseus's love for Penelope. For all our ideals about monogamy, we recognize that the death of an exclusive sexual bond need not mean the death of intense love.

Lest it seem that I've gone a little over the top in linking storks with Greek epics, consider that Bernd Heinrich, for one, doesn't shy away from attributing "love" to birds—or "grief," for that matter. He tells the story of Ruth O'Leary, an elderly woman from Idaho, who has had markedly close emotional ties with individual Canada geese. One goose, named Tinker Belle or TB, had been her companion for two years, even sleeping in Ruth's bed at night. At one point, TB flew off with a mate, and O'Leary felt sure she'd never see TB again. However, the next year, as she worked in her garden in the company of a young gosling, TB appeared suddenly with her mate.

Unsurprisingly, the gander hung back from human contact. TB, on the other hand, stepped right up into Ruth's lap, then followed her into the house, where she walked from room to room. In the bedroom, she pulled covers off the bed, perhaps assessing the best place to make a nest. In the living room, she pulled a videocassette off the shelf and looked directly at the television, which she and Ruth had watched together in the past. "Ruth," Heinrich writes, "took the correct tape, *Fly Away Home*, off the shelf and put it in the videocassette recorder. Tinker Belle leaped up onto the couch and watched more than half the movie—one she had watched often before." That evening, TB rejoined her mate in flight. A pattern was thus established. The goose pair would show up at Ruth's in the morning, TB would spend the day with Ruth, and the birds would fly off in the evening. Then one day, the gander was nowhere to be seen. For three days, TB flew around the area, calling and calling for her mate. After that, she sat with her bill under her wing and refused to eat. She became so weak that she staggered.

The gosling had remained at Ruth's house throughout this period. Once TB lost her mate, Ruth made sure that she interacted with the younger bird. The two swam and ate together, and at night both slept on Ruth's bed. Gradually, TB emerged from her sorrow and came back to her old self, eventually rejoining a wild flock. O'Leary attributes her recovery to the therapeutic effects of spending time with the younger goose. Here is an interesting echo of what happened with Willa, the cat who grieved for her lost sister Carson (as described in chapter 1), and whose spirits improved only when a younger cat came into the household.

Stories of bird affection fill Heinrich's book. Yet it's not as if attachment is so fixed in the birds' genes that whenever a male goose comes within courting distance of a female, the two tumble into a sort of preprogrammed rapture. Some couplings are perfunctory, the reproductive imperative carried out in the absence of anything that looks (at least to a human eye) remotely affectionate.

By contrast, Malena and Rodan behave toward each other in ways that may further their common reproductive goals but are not necessary for successful mating. The drive to produce offspring is fixed and the result of mating is inevitable, but the sharing of affection between any two birds is anything but.

And love between long-term partners? It's a trade-off seared in joy and pain. To love is to gain much—but also to lose when, after years and years, one is again, even if only for a while, alone.

While birds like storks, swans, and geese are linked in our minds with monogamy, the symbolic resonance of crows and ravens is more complicated. Corvids are birds of mystery and contradiction. They symbolize, on the one hand, trickery and deceit, death and doom. Yet they stand just as much for creativity, for healing and prophecy, and for the transformative power of death.

Notice how death figures on both sides of the corvids' symbolic power, the dark and the light. How could so much opposite meaning be invested by humans in one type of bird? In *Mind of the Raven*, Bernd Heinrich suggests that these contrasting themes emerged at different stages of human history. Ravens were revered, he says, when we as a species were hunters. Back then, where ravens flew, landed, and feasted, large animals could be found, animals whose meat sustained our lives as well. Later, when humans settled down and began to herd domesticated animals, the raven's association with death shifted. Now the raven became a thief, a stealer of that sustaining meat.

In some societies it was thought—as it still is by some groups—that ravens not only scavenged from animal carcasses but killed animals outright. This view is understandable, Heinrich notes; witnessing a raven plucking an eye from a dying calf would be reason enough to suspect the bird as murderous. In Yellowstone National Park in 1985, ravens were witnessed removing the eyes from a dying bison that was stuck in the mud, steam still puffing from its nostrils. It's the ravens' way to feast on corpses, and not only those of calves and bison. Humans too may become bird food. Historical accounts suggest that after great battles, when bodies were strewn across a field, ravens flew in to take advantage. This sort of behavior hasn't helped their ghoulish reputation.

It would be a stretch of the imagination to attribute the death of a bison or a person to a tiny bird, but severely aggressive acts by ravens toward other, smaller animals have been recorded. In the Arctic, a pair of ravens cooperated to kill seal pups resting on the ice. One raven would swoop down and land near a pup's ice hole. When the second bird drove the pup toward the hole, the first would peck the pup on the head until it died. Are these sorts of observations what led to the term for a social group of ravens, an "unkindness"? Still, that term is less harsh than the one for crows: a "murder."

With the advent of herding, in Heinrich's scheme, ravens' association with death became tinged with doom. An anthropological view, though, suggests no simple linear chronology of hunting first and herding later, but instead a dynamic and overlapping set of subsistence behaviors that responded to environmental conditions. Given this fact, perhaps disparate cultural traditions may be a primary cause of the raven's complex symbology. One sort of oral tradition regarding the raven may have emerged in one group, another sort in another location.

In their book *In the Company of Crows and Ravens*, John Marzluff and Tony Angell report a cornucopia of perspectives on corvids among Native American peoples. For Pacific Northwest native tribes, the raven may be seen as creator, clown, mischief maker, shape changer, or trickster. Some Native American groups relate a tale that explains the birds' jet color. The Lakota Sioux say that crows were at first white. When a Lakota hunter masked as a buffalo caught the crow leader, he threw the crow into the campfire in reprisal for the warning it had issued to other animals about the impending hunt. The crow escaped but was blackened by the fire. For the Acoma Indians, the crow became black for a different fire-related reason: after the crow created the world, he rescued it from fires; dipping his wings in water to cool the burning earth, he turned black.

Underlying all this complexity is the fact that the corvids are incredibly smart and social. As a primatologist, I love that they are dubbed "the feathered apes" based on their cognitive and behavioral similarities to chimpanzees. Corvid calls are not just expressions of fear or arousal but communicate specific messages about predators, family members, and available resources. People, say Marzluff and Angell, can easily determine corvids' emotional tenor by reading the "language" expressed through the varying arrangement of their contour feathers, an unusual communicative feature (and surely a key channel of communication among the birds themselves).

Corvids' social groups are hotbeds of shared learning and communication, where individuals are recognized and problems are solved intelligently. Sometimes it seems the birds gather with the express intent of sharing information with each other. At the University of Washington, crows gather each morning at a certain parking lot next to the football stadium. Flock after flock arrives in successive landings. Marzluff and Angell note the defeaning cacophony of calls, and even they, expert ornithologists, wonder what's going on. This parking-lot ritual has endured for forty years, with at least four generations of crows now involved. In the beginning, the location made sense in a practical way: a dump existed at the site, in which crows could scavenge for food. Now no food can be found nearby. The lot is not particularly warm, nor is it near the crows' roosting site. Why do the birds still select this spot to gather? The crows are doing what their parents did, and their grandparents. It's ritual now, a local tradition. The crows "regroup, catch up on the latest buzz, prepare for the day's events, and shake the sleep out of their bones," write Marzluff and Angell. It's a cultural choice the crows make.

The popular theme in human myth and legend of crows' and ravens' close association with death, then, can be assessed from a scientific point of view against the backdrop of these birds' social tendencies and intelligence. What we know about the feathered apes leads to a prediction that corvids may feel, and express, emotion when a flockmate dies. But *do* corvids really grieve? Marzluff and Angell report a curious thing that sometimes happens with crows: a loud, squawky cluster of birds, numbering in the hundreds or even thousands, gathers and stays together for about a quarter of an hour. Then there's a period of silence, followed by a collective departure. Left behind is a single dead crow. What could that be about?

Crows tend to avoid places of danger, including locations where humans or other predators have captured flockmates in the past. If the single crow has died right in the midst of the group, perhaps the strange silence accompanies a process by which the survivors fix in their minds a place to avoid in the future. But could something more emotional be going on, like a crow funeral?

A controlled experiment might help with this question. Marzluff and Angell reasoned that if they were to place dead crows in a known study area, they might evoke the peculiar noise-silence-departure sequence from the resident crows. Testing that hypothesis, they found that they could not in fact evoke that behavior, but what did transpire was revealing. Within minutes of the carcasses' appearance, resident birds uttered assembly vocalizations, which served to call in crows from the surrounding area. Soon, ten or more birds, all calling, circled over the dead ones. A few, residents of the study area, went to the ground for a closer look, perhaps checking to see if they knew the identity of the dead birds. A half hour later, it was all over. No silent period had ensued, and nothing that could be described as a funeral had taken place.

The parallels with the elephant research described in chapter 5 are notable. In Kenya, Cynthia Moss reported behavior that suggested a preference among elephants for the bones of their dead relatives. Specifically, she saw a seven-year-old male caressing, for a longer period than anyone else in his group, the bones of his mother's jaw. Such attention, if directed preferentially toward the bones of animals that were loved when alive, could be one measure of elephant mourning. Then Moss and two colleagues made the same choice that the corvid scientists did—to follow up on an impression by launching an experiment. It turned out that elephants did not prefer bones of their own matriarch to bones of other groups' matriarchs.

Just as I couldn't dismiss the meaning of Moss's anecdotal observation of an emotional elephant youngster because of the experimental data, I can't now dismiss the idea that something of emotional significance is going on with corvids' response to death. Marzluff and Angell don't dismiss it either. In their more recent book *Gifts of the Crow*, they devote a chapter to "passion, wrath, and grief" among corvids. There they recount what happened when a golf ball hit a crow on a Seattle golf course. The precision strike was an accident, of course; concerned golfers who witnessed the crow go down were startled to see that another crow immediately came to its aid. This second bird pulled on the first crow's wings, calling out all the while. Five more crows soon arrived. At that point, three crows together "began pecking and pulling on the apparently dead bird, trying to lift it up by the wings." The golfers assumed the bird would not survive and moved on, only to learn two holes later from other players that the stricken crow had in fact revived and flown off.

This anecdote sure looks like the expression of compassion for an injured flockmate. In some cases, though, corvids respond to an injured companion by killing it. And once in a while a crow mob will gang up and kill a crow who doesn't even seem to be injured. These are complex birds, and there's no predictable outcome of their social encounters. Still, the helping behavior described sets the stage for thinking about love and grief in corvids. Marzluff and Angell emphasize that crows and ravens "routinely" gather around bodies of their own dead. That response may be adaptive, they think, to the extent that it aids the birds in assessing what killed their dead confederate (thus increasing their chances of avoiding that fate themselves). It may, in addition, help the birds size up their new place in the flocks' shifting hierarchy.

"We also suspect," Marzluff and Angell write, "that mates and relatives mourn their loss." Given that we're talking about complex creatures—the feathered apes—I suspect it too.

9 SEA OF EMOTION
DOLPHINS, WHALES, AND TURTLES

In the Amvrakikos Gulf off the coast of Greece, a mother, alone in the water with her offspring, tried over and over again to revive her child. She lifted the small body above the surface, then pushed it down under the water, then repeated the cycle.

The mother was a bottlenose dolphin, and the calf was a dead newborn; the observers were scientists aboard a research vessel from the Tethys Institute. They found themselves watching "desperate behavior" on the part of the mother, who, over the course of two days in 2007, vocalized and touched her calf with her rostrum and pectoral fins. The mother seemed unable to accept the fact of her calf's death. Other dolphins from the mother's group, numbering about 150, occasionally approached to observe the drama, but they neither interfered nor lingered. It was just the mother, and the dead baby, together in the water.

The researchers felt concern for the mother, and with good reason. Over about four hours in the course of two days, they never saw her feed. Given dolphins' high metabolic rates, her total focus on the calf could have put her health in danger. The baby, meanwhile, had already begun to decay. Compared to the babies of monkeys and apes, marine mammal infants undergo rapid decomposition after death. The mother already had begun to remove pieces of skin and tissue from the corpse.

A passage from the scientists' report about this dolphin mother described their own compassion:

> The researchers on board did not feel like taking the calf away from the mother to perform scientific investigations (e.g., a necropsy of the calf). Their decision was intended as a form of respect towards a highly-evolved animal, the deep suffering of whom was obvious enough.

The researchers knew what they were seeing: maternal grief.

Just like monkey and ape mothers, dolphin mothers will sometimes carry the corpses of their dead infants. At times, though not in the Greek case, the mother's social group attends acutely to this behavior. An early report of this type, from 1994, concerns bottlenose dolphins off the Texas coast. Some fishermen noticed that one dolphin, probably the mother, was laboring to keep a smaller deceased dolphin from washing ashore. Several adult dolphins moved in a clockwise direction around the pair; when the fishermen moved in closer, the animals began to slap the water with their tails. The maternal behavior continued for two hours and was observed again the next day (it was assumed that the same mother-infant pair was involved each time).

In a separate anecdote from the same report, volunteers from the Texas Marine Mammal Stranding Network observed an adult dolphin pushing a dead calf in the water. When the volunteers' boat moved in, the adult swam away and resurfaced elsewhere, even as other dolphins swam nearby. By switching off the engine and drifting quietly closer, the rescuers managed to pluck the calf out of the water and bring it into the boat. At this, the mother became frantic, rushing underneath and at the boat.

I found myself wishing these volunteers had not intervened but instead had allowed the mother to experience her grief. The facts, though, temper this sentimental view: the mother could no longer help her baby, and she might have dangerously depleted her energy reserves had she kept on with such behavior. It's even possible that this second Texas observation of maternal behavior involved the same dolphins as the first: it occurred only six days after, and twelve miles distant from, the earlier sighting. Had the marine-mammal volunteers spotted the same mother that the fisherman had? Could the mother possibly have endured in her efforts for that long a time? That the infant in the second sighting looked to be in worse physical shape than the first suggests that she may have.

In 2001 off the Canary Islands, rough-toothed dolphins were observed by scientists over the course of six days during whale-watching trips. One animal, presumably the mother, was seen pushing a dead newborn in ways that are by now familiar. This time, though, the mother had escorts—two adult dolphins who, Fabian Ritter reported in the journal *Marine Mammal Science*, "were swimming in a highly synchronous way, usually slightly in front of and constantly escorting the mother." Other dolphins approached the mother and her escorts. The next day, similar behaviors were seen, and over the following four days, during four sightings, escorts were present three times. Meanwhile, the dead baby began to show signs of decay. By the fifth day, the mother left the calf alone for longer periods, but the dolphins still defended the corpse when a gull approached it. And, interestingly, the escorts began to participate in the calf-supporting behavior more directly: they swam, at times, with their backs underneath the body.

As is routine in the scientific literature, the published descriptions are dry, clinical, and stripped of emotion. But it may be significant for a profile of dolphin grief that adults other than the mother involved themselves with the infant's body. Working from their database of individual photographs, scientists identified nineteen dolphins who participated in some way in the unfolding events. Of these, fifteen were, on the basis of previous observation, thought to be members of a single, tight social unit. Further, the dolphins' travel speed was slower during this period than had ever before been observed. Over the six days, the dolphins moved very little from their original location. Ritter concludes that the group adjusted its behavior to the exceptional circumstances surrounding the calf's death.

A tightly coordinated response among animals who communicated closely with each other, this event shows clearly that a dolphin social group may be affected by an infant's death. It would be risky for me to assert that it shows as well the existence of collective dolphin grief. Yet the chain of logic isn't weak: Given that mother dolphins exhibit grief and that dolphin pods are tight-knit social groups, it is entirely possible that dolphins other than the mother exhibit grief when a youngster dies.

Gregarious creatures, dolphins play exuberantly with partners in their own groups and sometimes also with whales. A wonderful series of photographs taken in Hawaiian waters shows play between a bottlenose dolphin and a humpback whale, once near Maui and another time near Kauai. In both cases, the dolphins draped themselves across the whales' heads; the whales then reared up and the dolphins slid down their backs. At no time did the whales' actions appear aggressive, and the dolphins' cooperation over multiple "rides" was complete.

Would a dolphin mourn a whale play-partner who died, or vice versa? The play interactions, occurring outside of any long-term friendship, may be too fleeting for that. It may just be that these marine mammals are so primed to come together socially that their usual within-species play patterns spill over into cross-species play when an opportunity arises. The dolphins' expansive behavioral repertoire, surely rooted in emotion, leads me to think that a hypothesis of shared dolphin grief is highly plausible.

Whale-for-whale mourning may occur in relation to a phenomenon that concerns (and sometimes mystifies) marine-mammal scientists: mass strandings. In February 1998, 115 sperm whales beached themselves in three strandings on the coast of Tasmania. The individuals came from three separate groups, were mostly female (97 of the 112 whales who could be reliably sexed), and represented a variety of ages, from under one year to sixty-four years. In a paper for *Marine Mammal Science*, Karen Evans and her research team report valuable physiological details gleaned from the whales' carcasses. For instance, I was startled to learn that among the whales were pregnant females ranging in age from twenty-five to fifty-two years; I had not expected whales to be reproductively successful at such advanced ages.

At one of the three strandings, the whales' behavior could be closely monitored. First, a tight cluster of thirty-five whales moved from open waters toward the surf zone. One whale began to swim away from the others in a "frantic" way, churning up water and moving parallel to the shore, then stranding on the beach. In pairs and trios, the other whales followed the first to the surf zone; from there, wave action pulled them in to the beach. (The last two whales to strand deviated from this pattern; they swam past the others and actively beached themselves in a separate area.)

Sperm whales organize themselves into temporary aggregations of ten to thirty adult females and their offspring, with smaller permanent subgroups breaking away from the larger groups and rejoining them at various times. In their paper, Evans and her team put forward no connection between this family organization and the "why" of the strandings. When talking with the media, though, Evans noted that the whales probably beached because of a kind of emotional contagion: the original whale or whales became stranded for reasons related to distress or injury, and family members followed because they refused to abandon their kin.

This explanation for the sperm-whale stranding is tantalizing, and matches up with events in other whale species. When pilot whales strand, says Ingrid Visser of the Orca Research Trust in New Zealand, other pilot whales arrive to inspect what is going on; if rescuers try to herd them away, they become quite stubborn. "If we tried to get them to move past without stopping, they would fight to go back to the dead animal," Visser told the journal *New Scientist*. "I do not know if they understand death but they do certainly appear to grieve—based on their behaviors."

Dolphins strand in big numbers, too, and scientists agree there is no single factor that explains why. Between January 1 and March 7, 2012, 189 dolphins stranded on Cape Cod, Massachusetts, far in excess of the annual average of 38. One factor could be Cape Cod's hook shape, which may trap the dolphins in shallow waters—but a permanent feature of the landscape cannot account for a single year's spike in strandings. Nor can Cape Cod's topography explain the dolphin strandings that occur elsewhere. Causes of strandings, including the military's use of sonar, which may disorient the dolphins' navigational abilities, are hotly debated. In short, mass marine-mammal strandings are not well understood. Social ties, even social mourning, may help explain some of the whale and dolphin strandings, but these factors offer only a partial answer to a disturbing mystery.

So far, I've considered only cetaceans, but questions about mourning apply to nonmammals as well. Sea turtles are reptiles, and gorgeous ones at that. In their swimming grace, they seem wholly unlike the awkward-gaited land turtles with which most of us are more familiar. On the Hawaiian island of Oahu, a spot nicknamed Turtle Beach attracts numerous endangered sea turtles. Residents and visitors a few years back came to know and love a turtle they dubbed Honey Girl. Great sadness ensued when Honey Girl was found slaughtered (cruelly, by human hands) on the beach. Grieving residents set up a memorial to Honey Girl that featured a large photograph of her. Turtlelovers flooded the memorial, but an unexpected visitor showed up too. A large male sea turtle hauled himself out of the water and made his way up the beach straight toward the photograph. There he parked himself, in the sand, head oriented toward the image of Honey Girl. Judging a turtle's gaze as best humans can, observers concluded that he stared hard at the picture for hours.

Was the male grieving for his mate? All along, we have considered how we might come to discern a wild animal's emotions; doesn't this question only increase in complexity when dealing with a reptile? A turtle is, after all, many evolutionary eons away from us primates, and indeed from any mammal—it is a creature cold in the bone to our hot in the blood, as psychologist Anthony Rose puts it. When we posit that a turtle is grief-stricken (as televised news reports did in the case of Honey Girl's presumed mate), aren't we imposing romanticized notions upon a species that operates on instinct?

We will never know with certainty that Honey Girl's mate mourned her on the beach, or even that he knew the photographic image depicted Honey Girl. Clues do suggest that something was going on in the male's mind, something more than a mere attraction to novelty on the beach. His straight-arrow path to the memorial, and the quality of his stillness during his hours in front of it, are notable. Would he have behaved the same way had he encountered a sand sculpture of Honey Girl roughly the same size as the photo, or some other large novel object unrelated to Honey Girl? Short of jetting to Oahu to run a controlled experiment, I cannot say for sure. Whatever that turtle was up to at Honey Girl's memorial, though, it's clear to me that he was acting out of choice, behaving in a realm that stretches beyond mere survival activity.

My own experience with tortoises, and turtles, emerges from less exotic locales. I regularly encounter them on roadways as they amble across lanes of traffic, unaware of the imminent risk of becoming brightly colored bits of roadkill. Turtle rescue gives me a thrill, I admit: a quick carry from midroad to the safer verge for the smaller, amiable ones, a behind-the-shell foot-shuffle to guide the bigger, hissy ones (while avoiding their snap-rapid jaws). One summer day, after a quickly executed pullover onto a highway's shoulder, I joined in epic battle with a magnificent snapper poised on the edge of trouble. Plucking the creature from the path of predatory vehicles, I set her (or him) down on grass, and reoriented her toward safer pastures. Back she wheeled, heading once more for the thick ribbon of cars. Perhaps seeking a watery oasis across the road, and thus set to "instinct," she resisted all aid. Finally, carrying her aloft, I plunged through the smelly and brackish roadside ditch water (sacrificing clean sneakers and pride, as passing drivers gaped) and placed her out of harm's way. Self-contained, methodical, stoic: that's a turtle's nature. "Eat, Move, Mate" would be the turtle world's best-selling book and movie title. Wouldn't it? So I once assumed. But, applying the questions raised by the Honey Girl anecdote, I now think it lacks rigor to assume a single "turtle nature." Tortoises and turtles, I'm learning, not only come in diverse species and sizes, on land and on sea, but behave in ways that go beyond the instinctual.

Consider the tortoise who aimed to help a companion in distress. Here again we benefit from the craze for videotaping the actions of any animal that is cute, comic, or doing something unexpected. In this clip, a tortoise lies canted on its side, legs angled uselessly to the sky and unable to right himself (or herself). A second tortoise approaches. Tortoise B pushes his face right up near A's body, perhaps to assess the situation, then begins gently to push on A. Nothing much happens at first, but B continues to labor with purpose and precision. Once A begins to tilt back toward the ground, he wheels his legs, thus adding his own force to B's. When A regains his quadruped stance, the pair moves off together, slowly. With a video of unknown origin like this one, it's possible that viewers, including me, have been suckered. Could Tortoise A have been placed on his side by a person eager to offer a dramatic scene to a YouTube-addicted world? And what about ethics? Shouldn't the videographer have helped Tortoise A early on, even before Tortoise B stepped in? Even though the circumstances surrounding this video aren't clear, the inventive—and successful—problem-solving behaviors shown by Tortoise B are striking.

As I remarked in writing about goats and chickens in the prologue, what we notice in the animals around us is set, to some significant degree, by our expectations. We may not even think to look for mourning when a turtle lose a partner. We may not think to look at turtle behavior very closely at all. Yet to be ruled by our assumptions leads to missed opportunities, a lesson brought home by Verlyn Klinkenborg's fictional tortoise in his novel *Timothy, or Notes of an Abject Reptile*. Timothy was born among the tangy salt smells of Turkey and transported to England on a ship. What Klinkenborg reveals, through Timothy, is that we humans don't understand other animals nearly as well as we like to think.

Timothy offers an ethnography of sorts, a view of the *Homo sapiens* who weathered the eighteenth-century English winter in ways that are peculiar to Timothy's sensibility: "Humans of Selborne wake all winter. Above ground, eating and eating. . . . Huddled close to their fires. Fanning the ashes. Guarding the spark. Never a lasting silence for them. Never more than a one-night rest." Reflecting further on the human condition, Timothy finds little to envy: "Barely able to witness what is not human. Always conjuring with the separateness of their species. Separate creation. Special dominion. Embarrassed by signs of their animal nature." More than anything, Timothy is flummoxed by humans' drive to measure, categorize, and rigidly label the natural world, all the while puffed up with the resolute certainty of their understanding. All through his notes and descriptions, the human Gilbert White (a real-life English naturalist of the eighteenth century who wrote about tortoises) refers to Timothy as "he." White has never seen any evidence to suggest that Timothy is anything other than male, so he leaps to a conclusion. "No eggs buried under the monk's rhubarb," Timothy reflects, "or hidden at the foot of the muscadine vine. None laid on the grass-plot. No preening, no dalliance. . . . And so Mr. White has always supposed that I am male."

As Klinkenborg's narrative reveals, Timothy isn't male. She is full of surprises, both about her sex and about her ways of living in the world. I am drawn to this novel because it mirrors perfectly what we are coming to grasp more acutely in animal-behavior science more clearly than ever before: We must look at animals' actions with fresh eyes and thoughts unconstrained by expectations.

When in 1994 animal behaviorist Gordon Burghardt visited the National Zoological Park in Washington, DC, he stopped at the enclosure of a Nile soft-shelled turtle named Pigface. Enclosed alone, Pigface had by that time lived at the zoo for fifty years. (Reading that statistic, I had to stop for a moment and let it sink in: five decades captive.) Burghardt had looked at Pigface before, but this time he did a double take: Pigface was playing with a basketball. The turtle swam through the water, batting the ball with his (or her) nose and chasing it with great energy. This snapshot of turtle play invited Burghardt to think in a new way about the behavioral repertoire of reptiles.

In the twenty-first century, we tend to veer between two poles in thinking about creatures of Pigface's ilk. We may conclude that the male Hawaiian sea turtle was mourning his mate, Honey Girl, or we may look at turtles and tortoises much like the fictional Gilbert White does, boxed in by assumptions that their lives are circumscribed by the "eat, move, mate" circuit. I don't think that the Honey Girl anecdote proves the existence of turtle grief, but like Pigface and his play behavior did for Gordon Burghardt, it should shake us into a realization: We won't have a hope of finding turtle grief until we look for it.

10 NO BOUNDARIES
CROSS-SPECIES GRIEF

The bulky gray body, with its huge ears and dangling trunk, walked in a big open field beside a smaller, white romping one. Tarra and Bella were out for a walk. Side by side, day after day, they roamed the open acres of the Elephant Sanctuary in Tennessee. They even took swims together. The trust that Bella the dog felt for her friend was evident when she allowed Tarra to caress her stomach with one massive foot.

Tarra bonded with the stray dog Bella all on her own, without any urging from her human caretakers. For eight years, the two were fast friends. And thanks to television and the Internet, they became a global video sensation. That two creatures of such disparate size, indeed of such different natures, shared an enduring friendship was uplifting news for many people. Tarra and Bella remind us that when individuals will it so, bonds of friendship may transcend even extreme dissimilarities.

Then one day in 2011, Bella was attacked by a wild animal, or possibly more than one. The attackers were almost certainly coyotes, and they killed her. Though circumstantial, the evidence that could be gathered points to two conclusions: Tarra was the first to discover Bella's body, and she carried her dead friend back near the barn where the two had spent happy times. No person at the sanctuary witnessed Tarra's discovery or carrying of Bella's body, so I cannot affirm the truth of these conclusions, but here are the known facts: Tarra and Bella were seen together on October 24, 2011. The next morning, and indeed through the day, Bella was nowhere to be found. Sanctuary workers began a search for her that yielded no results, and continued it the next day. Bella's

TARRA AND BELLA AT THE ELEPHANT SANCTUARY. © *THE ELEPHANT SANCTUARY IN TENNESSEE.*

prolonged absence was so unusual that people at the sanctuary began to fear the worst.

And then those fears were realized: Bella's body was found near the barn. There was no sign of coyotes or any other wild animals, or of any altercation, visible near the body. How she got there is a puzzle. Bella might have wanted to make her way from the site of the attack back toward the barn, a place of comfort for her, yet her injuries were probably too severe for her to have traveled that distance on her own. When the sanctuary staff discovered blood on the underside of Tarra's trunk, they concluded that Tarra had carried Bella back to the barn. Or perhaps Tarra discovered Bella making her way to the barn, or found her there, at the spot where she would die, and offered her help or comfort with her trunk.

In any case, Tarra showed no interest in lingering with Bella's body once caretakers brought her to it. Later that day, when the little dog was buried, Tarra did not approach the ceremony. On its website, the sanctuary later reported the events of that and the following day:

> Tarra chose not to participate in her burial. She was close, less than 100 yards away, on the other side of some trees but she would not come over. She had already said goodbye. This was for the humans. . . . The following day, caregivers made the heartbreaking discovery that Tarra had gone to visit Bella's grave sometime during the night or early morning. They found fresh dung nearby and an elephant footprint directly on Bella's grave.

At first, my response to this last claim was a skeptical one. How could the identity of the elephant visitor to Bella's resting place be known? From sanctuary caretakers, I learned some key details. While Tarra had not been observed directly at the grave, she had been seen in its vicinity—and no other elephants had. Further, skilled observers can discern an elephant's identity from footprints and dung alone. It was these factors taken together that led sanctuary staff to conclude that it was Tarra who had visited Bella's grave.

What is beyond question is how much joy Tarra and Bella had each taken in their long-term friendship. Here is a situation in which a response of grief would be highly predictable for the surviving partner. Yet one final detail of the sanctuary's report deserves attention. After Bella went missing and before her body was found, Tarra's caretakers already judged her to be depressed and grieving. The elephant ate less and behaved in atypical ways. Because of the timing, Tarra was at that point upset about an *absence*, not a known death. We have grappled with this distinction before: How to distinguish an animal's emotional response when she cannot locate a friend from a state of outright mourning?

It's common enough for a zoo elephant, gorilla, or chimpanzee to discover one day that a close friend of many years is simply . . . gone. The friend may have been crated up and transported to another zoo, with no way for caretakers to explain that fact. And this isn't so different a situation from what may unfold in our own homes, when a pet dies at the vet's office and a friend is left back home. Even in a sanctuary setting, it takes an insightful observer to distinguish between an elephant who pines for playful companionship in the immediate moment and an elephant who feels grief. In Tarra's case, what started out as in-the-moment sadness seems to have blossomed into full-on mourning. Her caretakers reported that Tarra continued to visit Bella's grave on and off for weeks after the dog's death.

The depth of feeling between Tarra and Bella helps explain why, in the last few years, cross-species animal friendship has become a wildly popular topic. Tarra and Bella played a role in this phenomenon when video clips of their friendly interactions went viral on the Internet. Then in 2011, Jennifer Holland of the National Geographic Society published *Unlikely Friendships*, and it hit the best-seller lists. Friendships between a sled dog and a polar bear, a snake and a hamster, and forty-five other pairs—including Tarra and Bella—are explored in her book. Holland describes Tarra's earlier vigil when Bella was ill; in seeming distress, Tarra waited for many days outside the house where Bella was being nursed back to health. When the two were finally reunited, each expressed joy according to its species: Bella wiggled her whole body and rolled on the ground; Tarra trumpeted and stroked Bella with her trunk.

Sometimes, what gets labeled as a cross-species friendship is more accurately described as a short-term positive association. Think of it this way: You are a guest at a friend's house for several days, and enjoy backyard romps with your friend's dog. You initiate the first bout by throwing a Frisbee, but later, the dog invites you to play by bringing you a throw toy; from his body language, you know he's having fun. Each of you is a willing partner in a series of positive interactions, a kind of temporary alliance that fits the circumstances of the moment. But did you and the dog forge a friendship? Only if the criteria for defining a friendship are satisfied by fairly fleeting interactions.

But *should* lengthy association be a requirement for friendship? In *Unlikely Friendships*, Holland tells of a sled dog and a polar bear in the northern Canadian town of Churchill. One day a large bear approached an open corral where sled dogs were chained. Common in the area, wild bears sometimes kill sled dogs. Although most of the dogs responded anxiously, one did not. Photographer Norbert Rosing watched as the bear rolled over and stretched out a paw toward that dog. Cautious at first, the dog began to relax, and respond to the bear's play invitation. At one point, the dog cried out in pain when the bear bit hard, but from that point forward the bear checked his strength in deference to his smaller partner. The play bout lasted about twenty minutes, and the bear returned over the next several days to play with the dog.

In Churchill, this kind of play is not unique to a single bear-dog pair. Multiple bears may romp with multiple dogs in a sort of cross-species playfest. A video clip captures the massive bears, dirty-white against the whiter snow, using slow-motion movements that engage but don't frighten the dogs. One bear nudges a dog with his big blunt snout; another folds a dog in a literal bear hug, causing the dog to squirm. Once, a bear opens his jaws right around a dog's head. Yet the dogs are relaxed around the bears and come back for more.

The bear-dog play behavior cries out for more study. Are the play pairings random, that is, will any bear play with any dog? Or do specific partners choose each other time and again? What happens when the bears and dogs are apart for some time? Is there any indication that one partner misses the other? Has a bear ever encountered the carcass of his dog play partner, or a dog come upon the body of his bear play partner? Is there anything approaching a Tarra-Bella relationship among the dogs and bears of Churchill, such that cross-species grief might follow when a play partner dies?

Not all cross-species friendships are as ripe for questions about mourning. Take, for instance, the bond supposedly shared by a snake and a hamster. The hamster was introduced to a zoo snake in winter, when the reptile's metabolism was low, and the snake cradled the hamster in its coils. Holland admits that had the meeting come about in summer, the outcome might have been a rodent-shaped lump in the snake's body. What happened to the hamster as time went on? Holland doesn't say, and I wouldn't anticipate any evidence of animal grief emerging from this scenario.

The friendship between the hippo Owen and the tortoise Mzee is, on the other hand, remarkable for its constancy. Orphaned during the terrible Christmas 2004 tsunami, Owen was brought to the Kenyan animal park where 130-year-old Mzee lived. Although no dramatic spark flared up between the two, the pair, led by the younger, rambunctious Owen, gradually developed a shared affection. Before long, each followed the other around and an idiosyncratic communication system emerged. Mzee nips Owen's tail to propel Owen along on a walk. Owen nudges

Mzee's feet when it's his turn to initiate: he pushes on Mzee's back right foot when he wants Mzee to steer right and does the opposite for going left. What will happen when Owen loses Mzee, or Mzee loses Owen? The price of an enduring friendship is often survivor's grief, and we know that grief does not respect species' boundaries.

Cross-species friendships, and the grief that may follow, may be found in our homes as well. Melissa Kohout was moved by her cat Madison's response to the death of her dog, a Doberman called Lucie. Madison had joined the family as a kitten when Lucie was four years old. Because Madison arrived with ringworm, she required a bath every evening for weeks, and Lucie took it upon herself to lick the kitten dry. For years, the two animals groomed each other each night. Seven years along in this relationship, Lucie the dog fell ill with cancer. During this difficult time, a funny incident occurred. As cats will do when "gifting" favored humans, Madison brought a rat into the bedroom late one night and dropped it on Kohout's chest. "Covers went flying," Kohout told me, "and cat and rat ran into the kitchen. The rat bit Madison on the front paw, and she screamed. Lucie, as sick as she was, came running, bit the rat in half, and went back to bed."

When Lucie died, it happened at home. Madison climbed into the bed and burrowed under the covers, something she had never before done. For about the next month, she emerged from that self-made cave only to eat and use the litter box. After that "time of mourning," as Kohout puts it, Madison never again sheltered herself in the bed in that way.

Karen Schomburg describes an instance of cross-species grief that occurred on her small farm in the state of Washington. At thirty-two years of age, her Shetland pony Peaches fell ill with shortness of breath and congestion. Jezebel, a goat who had been friends with Peaches for years, showed great concern, refusing the company of other goats in preference to time spent with Peaches. Worried enough herself to make frequent checks on the pony, Schomburg saw something surprising late one night: Peaches had backed up against the manger to steady herself as she grew weaker, and Jezebel was pressed up close to her, leaning into her chest. Only with the extra strength from her friend could Peaches stay on her feet. In the morning, however, Peaches was on the ground, dead. To Schomburg, Jezebel looked forlorn.

The story of Peaches and Jezebel shows that, even when animals have their own kind around them (unlike Owen and Mzee), they may opt for a cross-species friendship. With other goats available, why did Jezebel seek the company of a horse? Why did Tarra, surrounded by other elephants at the Tennessee sanctuary, desire Bella's canine company? How did these cross-species friendships come to matter so much that the survivor became emotionally involved when the friend was dying (as with Jezebel) or after she had died (as with Tarra)? Many animals are curious, sociable, and open to new experiences. They may seek out more of the "positive vibes" that come from an initial interaction with another creature, and a friendship may be the result.

In a way, cross-species mourning is woven through many of the stories in this book. Animals may grieve for a human companion who dies, and we may mourn the animals we love and lose. Berlin, Germany, underwent a citywide outpouring of grief for the polar bear Knut, who died in 2011. Knut became a "national obsession," as the *New York Times* put it, when he thrived in the Berlin Zoo even after rejection by his mother. In three places—the neighborhood of Spandau, the National History Museum, and the zoo itself—memorials have been or will be erected in honor of the bear.

An urge to memorialize occurs, too, on a smaller scale. Recently I attended a brief ceremony after the death of a cat named Tinky. For eighteen constant years, Tinky had been the companion of my friend Nuala Galbari, who with her partner David Justis cares for a variety of animals, from cats and rabbits to horses and birds. When Tinky was a kitten, Nuala played the piano with him next to her on the bench; she fell into the habit of moving his paws gently over the keys. Tinky not only responded positively but began to play musical notes to communicate with Nuala. When Nuala developed a debilitating illness, Tinky, attuned to her weakened state, stayed by her bedside. During her long recovery, the bond between the two was cemented. When she was again healthy, Nuala continued to share a love of music with her cat. "On several occasions," Nuala says, "Tinky played up to six notes in an octave with his right paw. Following applause, he might then decide to play some lower notes with his left paw. No doubt, the little cat had somehow figured out that I played with both hands, and so he used both paws." When Tinky, much older and in a weakened state, took his last breath and died at home, a small group of us felt his loss keenly. We came together where he was buried, in Nuala and David's backyard, to share photographs and poems that evoked Tinky's life.

In *The Last Walk: Reflection on Our Pets at the End of Their Lives*, Jessica Pierce writes movingly of the last weeks, and the death, of her dog Ody. Ody was a Viszla breed, fourteen years old. In his extreme old age, his legs were severely atrophied, he had dementia, and he was almost completely blind and deaf. Pierce, a bioethicist, worked to figure out what a "good death" would mean for Ody: what she owed Ody, what time and manner of death was right for *him* and not only for her own fierce attachment to him. Of course, Ody wasn't always infirm. For many years, he had been Pierce's running and mountain-biking partner. Even now, with his ill health, she worried that what seemed to her a terribly diminished life didn't seem so to Ody. But things were getting worse; Ody was falling and, unable to get up, lying in his own poop until someone in the family found him. Eventually, Pierce arranged for a vet to come to her home and euthanize Ody.

After Ody died, Pierce realized that, wrenching as it was, the moment of euthanasia wasn't the most difficult:

> For me, the anticipatory grief—the sense of impending loss—was by far the worst stage. I mourned for Ody long before he even came close to death. The moment of his death was sharp and painful—the kind of grief that makes you feel as if you're drowning. But that didn't last more than a few hours.

What I like best about Pierce's writing is its honesty. She describes Ody as "one of my greatest loves and also my millstone, for fourteen long years." Ody wasn't always in poor health, but he was always a high-maintenance sort of dog—contrary and neurotic, in Pierce's words. I understand the love, and I understand the millstone comment too.

In the den of my home, on the mantel, sit eight small cherrywood boxes. Five contain the ashes of family cats, two of family rabbits, and one, the biggest, the cremated remains of a dog we cared for after my brother-in-law died. On each box is affixed a small plaque with our chosen words of memory. For feisty Gray and White, originally a proud and distant big-tom feral cat who fell in love with indoor living after he became ill and we took him in, the plaque reads: "Alpha feral, who finally found the love he always deserved." For quiet Michael, who lived only three years and also battled a number of complicated medical conditions, something pithier: "The sweetest boy."

Some of our lost animals were high-maintenance. Some were contrary, and neurotic. We grieve for them all, who were our friends.

11 ANIMAL SUICIDE?

Bear farm: It's a term that jolts. Chicken farm, cow farm, pig farm, even bison farm or llama farm, these places are familiar. Once past childhood's innocence, we carry images of farm animals that aren't always bucolic; that animals may be killed in ways far from humane is something we know. Sometimes an extra layer of information brings the situation into terrible focus; Annie Potts' description in her book *Chicken* of what occurs in chicken slaughterhouses haunts me, too much so to repeat the details here. However we respond to such knowledge—whether we adopt a vegetarian or vegan diet, or select free-range meat from local farms, or eat meat from any available source- the farming of chickens, cows, and pigs is a familiar practice.

Bear farming is not; at least it wasn't for me until quite recently, when one of Marc Bekoff's blog posts caught my eye: "Bear Kills Son and Herself on a Chinese Bear Farm." After a bit, my mind shifted from taking in the basic concept of a "bear farm" to considering the headline's main point. Bear suicide? In August 2011, first Chinese and then some Western media published accounts of the incident to which Bekoff refers. The UK's *Daily Mail* online, not known for journalistic circumspection, declared in its own headline, "The ultimate sacrifice: Mother bear kills her cub and then herself to save her from a life of torture."

In order to reflect clearly upon this bear's actions, and their possible relationship to bear grief, I need briefly to tackle a disturbing topic: what happens at a bear farm. More correctly, it's called a bile farm. Across Asia, from China to Vietnam and South Korea, bears are held captive because their bile contains a compound that is considered medically valuable. This substance, called ursodeoxycholic acid or UDCA, is touted as useful in fighting liver disease, high fevers, and other ailments. Further, some companies put bear bile into products like gum, toothpaste, and face cream.

The biology of this situation is complicated, because it's not only bears that produce the sought-after bile. Many animals, including humans, produce it as well, and a synthetic compound called ursodiol has been created and used to treat gallstones. Rather than signaling the end of bile farms, however, these alternatives have apparently pushed things in exactly the wrong direction. Else Poulsen writes in her book *Smiling Bears* that the appearance of alternatives not dependent on bear bile has backfired for bear welfare, as their artificial nature has lent cachet to the genuine article. The bile extracted from living bears has become an expensive trophy among a certain moneyed set.

Poulsen's book offers a heartbreaking tutorial about what goes on at bear farms. (This is a paragraph to skip if you wish to avoid vivid descriptions of animal suffering.) In China, Asian black bears become nothing more than living bile machines. "Each bear," Poulsen writes, "lies down, permanently, in a coffin-shaped, wire mesh crate for his entire life—years—able to move only one arm so that he can reach out for food." "Permanently" is the searing word, and it comes up again in another of Poulsen's passages: "Without proper anesthetic, drugged only half-unconscious, the bear is tied down by ropes, and a metal catheter, which eventually rusts, is permanently stuck through his abdomen into his gall bladder." Over time, some bears simply lose their wits. Unable to free themselves, they bang their heads on the bars; the relief of death comes far too slowly.

Estimates vary for the number of bears held captive at bile farms across Asia, but the number seems certain to exceed ten thousand. One of these captives was the mother highlighted by Bekoff. The sequence of events seems to go like this: As a worker at the farm prepared to harvest his bile, the cub cried out in distress. Somehow, the mother broke free, grabbed her cub, and hugged him with such power that he died of strangulation. Then she ran headfirst into a wall, and died.

That description, of course, is far from adequate. Important information is missing. What exactly was the worker doing? How did the mother break free? Just as significantly, I've stripped the mother of any intention, motivation, or emotion. That's how I was taught to write about animal behavior in graduate school, but (as this book attests) that's not how I write about animals anymore. Here's an alternative version: The cub cried out in distress as a worker prepared to harvest his bile. The mother, distressed by her loved infant's pain, broke free, and squeezed the life out of her baby so that he would no longer suffer. Overcome by her own emotional pain, she ran, purposefully, headfirst into a wall, killing herself.

Which account is the more accurate? In the first place, it's hard to focus on such an analytical question when sorrow for these two bears, and thousands more, hits so hard. The underlying scientific questions are important, though: Could the mother bear have gone insane—an outcome some of Poulsen's passages suggest is possible—and flung herself toward the wall without any sense of what she was doing? Can some animals make a conscious choice to kill themselves? A witness quoted in a newspaper report made the claim that the mother killed her cub "to save it from a life of hell." Can some animals reason to the degree necessary to justify such an assertion? Can bears carry out what is in effect a mercy killing? We know that the flip side of love is sorrow, and the flip side of shared joy is solitary grief. Can sorrow go so deep that it causes an animal to bring about a loved one's death in order to free him from physical suffering?

Unfortunately, details of the mother bear's behavior are too sketchy to allow firm conclusions even about what exactly happened, and in any case, there's no way from observation alone to figure out *why* the mother bear did what she did. But let us not allow her and her cub to be filed away only as an unsolved mystery. Instead, let us use their fate—and the mother's behavior, whatever her underlying intentions—to add new questions to the others we have asked about animal grief. Do animals kill themselves? And if they do, is grief ever the probable motivation?

In 1847—a dozen years before Charles Darwin published *On the Origin of Species* with its theory of evolution by natural selection—that question was alluded to in the pages of *Scientific American*. The animal under consideration was a gazelle in Malta, but in some ways the story parallels what we know about the mother bear in China. Here is the brief item, published over 160 years ago under the headline "Suicide by a Gazelle":

> A curious instance of affection in the animal, which ended fatally, took place last week, at the country residence of Baron Gauci, at Malta. A female gazelle having suddenly died from something it had eaten, the male stood over the dead body of his mate, butting every one who attempted to touch it, then, suddenly making a spring, struck his head against a wall, and fell dead at the side of his companion.

The female gazelle's death came from natural causes, which sets her story apart from the bears'. But then there's that strange coincidence: the male gazelle, just like the female bear, slammed himself into a wall. Did these two creatures act in such dramatic and fatal ways because their emotions were overwhelmed by loss (for the gazelle) and suffering (for the bear)? Revisiting in 2011 the brief note about the gazelles, *Scientific American* blogger Mary Karmelek found grief-crazed suicide to be an improbable explanation for the gazelle's actions. She speculates about other explanations for his fatal behavior. Perhaps the male ate the same food that caused the female's death, but in his case it led to neurological damage that caused him to run amok. Or maybe it was stotting gone awry. Stotting occurs when, in fleeing a predator, a gazelle leaps into the air so that all four legs are off the ground at once. "What seemed a suicide," Karmelek writes, "may have been the male gazelle's unfortunately timed response to perceived human predators."

Bear suicide, gazelle suicide . . . there's reasonable doubt in each case. Both times the animal acted quickly, on what has been interpreted as a spontaneous impulse to die. Rash behavior of this sort is a staple feature of the anecdotal accounts that pop up when one enters "animal suicide" as a search term into Google. The idea of animals killing themselves seems to be an attractive one, in a strange sense, perhaps because it's one more way to recognize animal emotion and to feel a kinship with other creatures. Yet a good percentage of the time the "suicide" label is clearly inaccurate.

The classic example of an animal suicide myth involves lemmings. We've all heard that cliché deployed to characterize someone's behavior. Seeing a friend conform to some trend we find distasteful, we may admonish her to stop following the herd: "Don't be such a lemming. Think for yourself!" Lemming conformity is rooted in the idea that these small rodents plunge en masse from cliffs, each following its predecessor over the edge to its death. There's a two-part explanation for how all this fanciful notion got started, however, and it has nothing to do with suicide.

Part one involves the species' natural behavior. Lemming populations tend to fluctuate to a degree that may be significant. When the population density shoots up, some lemmings may migrate to avoid the intense competition for resources in their home area. It is true that large numbers of lemmings move around in herdlike ways—they just don't jump off cliffs. That second, crucial ingredient of the myth was supplied by Hollywood, as Australia's *ABC Science* explains. In 1958 the Walt Disney studio released a movie called *Wild Wilderness*. In the making of the film, lemmings had been needed, but because none resided on location in Alberta, Canada, the filmmakers purchased some from Inuit children in the area. *ABC Science* reports, "The migration sequence was filmed by placing the lemmings on a spinning turntable that was covered with snow, and then shooting it from many different angles. The cliff-death-plunge sequence was done by herding the lemmings over a small cliff into a river." Thankfully, no such calculated exploitation of animals could happen in the present US film industry. That segment of the Disney film became famous in its day, however, and through it the lemming legend was born.

The lemming example is an interesting one for a discussion of animal intentionality, because the mass suicide (in the myth) is seen as mindless, collective behavior. The idea is not that each individual lemming wishes to die and acts accordingly—just the opposite. Most of the lemmings have no clue what the lead lemming is doing, and so all perish together. Inevitably, questions of definition arise, just as they have with animal love and grief. Should the term "animal suicide" be restricted to cases where an animal acted through conscious choice to end his or her life? In the cases of the mother bear and the male gazelle, this restriction wouldn't help us much. In neither case do we know whether conscious choice was involved. But lemming "suicide" would be ruled out, and perhaps the definition might aid in excluding other candidate cases of animal suicide as well.

Near Dumbarton in Scotland, one can visit a place that locals have dubbed "the dog suicide bridge." During the last half century or so, over six hundred dogs have fallen from the Overton Bridge to their deaths. With mentions of "suicidal dogs" and "kamikaze canines," the media sensationalizes the situation by inferring that the dogs are purposefully jumping to their deaths. It strains credulity, though, to think that hundreds of dogs (one by one, not in a group as with the mythic lemmings) might choose to kill themselves at this place (or any other). So what is going on?

Probably the dogs' perception is in some way involved. The dogs may smell a prey animal, which they begin to track while high atop the bridge. As photographs attest, the architecture is such that dogs walking across the bridge would be unaware of the large drop-off on either side; from a dog's-eye perspective, only a low wall is visible. The leaping dogs are victims, it seems, of an unfortunate collision of architectural design and the biology of their own perception. No conscious intention for suicide need be invoked. The Scottish dogs are an open-and-shut case.

Still, might some sentient animals feel such emotional pain that they would act on an intention for suicide? The mammal-behavior expert and trainer Richard O'Barry swears that he saw a dolphin choose to kill herself, right in front of his eyes. The dolphin was Kathy, one of the cetacean stars of the 1960s television show *Flipper* that I loved as a child. According to O'Barry, Kathy locked eyes with him, sank to the bottom of her tank, and stopped breathing. "The [animal entertainment] industry doesn't want people to think dolphins are capable of suicide," he told *Time* magazine in 2010, "but these are self-aware creatures with a brain larger than a human brain. If life becomes so unbearable, they just don't take the next breath. It's suicide."

Time featured O'Barry's recollection in a story about the film *The Cove*, named best documentary of 2009 at the Oscars. Directed by Louis Psihoyos, the movie tells the story of O'Barry's activism against a heinous practice that occurs yearly in the small Japanese town of Taijii: the killing of thousands of dolphins, six months out of every twelve. (I haven't been able to bring myself to watch *The Cove*, because in some scenes the camera is turned on dolphins who undergo slow, agonizing deaths.) This brutal practice, driven by the lucrative business of falsely marketing dolphin meat as whale meat, remained largely secret before *The Cove* hit it big. O'Barry stresses that the vast majority of Japanese people had been as unaware of it as everyone else. The cove's location is secluded, and the dolphins' killers were highly motivated to keep their activity quiet.

It was O'Barry's history with dolphins that led him to animal activism and to his goal of exposing the Japanese dolphin slaughter to a wide public. Back in the 1960s, O'Barry had captured five dolphins from the wild and trained them to perform in *Flipper*. Once the five were installed in the Miami Seaquarium, O'Barry spent countless hours in their company. Once the show began to air, he and the dolphins watched it together on a television set brought right to the water's edge, every Friday night at 7:30. That's when O'Barry first realized that dolphins are self-aware: the dolphins—Kathy included—recognized themselves on the small screen.

In support of his claim that Kathy committed suicide in her tank, O'Barry points to the way dolphins breathe. For humans, breathing is an automatic process that requires no conscious thought. We breathe naturally, even while in deep sleep, and rarely think about our breathing during the day, except in special circumstances such as hard exercise or a moment of emotional upset. As I type at the computer right now, I am intently focused on choosing the right words to convey my ideas; I inhale and exhale without awareness of doing so. As "conscious breathers," however, dolphins enjoy no such luxury; they must focus on the drawing of each breath. According to O'Barry, when a physically healthy dolphin chooses not to breathe, she intends to bring about her own death.

When twenty-six dolphins died off the coast of Cornwall, England, in summer 2008, one expert suggested suicide as a possible explanation. The dolphins beached themselves in four separate spots on a river in south Cornwall. When word of the stranding first got out, rescuers rushed to the scene and managed to save perhaps ten to fourteen others (in the immediate frenzy, no good census seems to have been taken). For reasons no one understands, the dolphins who died had ingested large amounts of mud; their lungs and stomach were simply full of the stuff.

Notably, no fish were found in the dolphins' stomachs, so the idea that the creatures stranded while foraging for fish was ruled out.

In reporting the mass death, England's *Guardian* newspaper quoted a pathologist who examined the animals on behalf of the Zoological Society of London. Vic Simpson told the reporter, "On the face of it, it looks like some sort of mass suicide. We have seen strandings on beaches, sometimes with five to seven dolphins—but never on a scale like this." O'Barry, then, isn't a rogue voice in claiming the possibility of suicide by dolphins.

But what could be the dolphins' motivation to strand? Unlike Kathy, kept in captivity by people in the entertainment industry, here were healthy animals (a fact confirmed by autopsy) swimming free in the wild. As it turns out, the British Royal Navy had been conducting sonar exercises in the area at the time of the dolphins' death. The Ministry of Defence was quick to say that these exercises were too far away to upset the dolphins, but perhaps this remains an open question. Could the sonar have caused the dolphins to become confused and panic? Either way, the suicide hypothesis seems to me not to be ruled out by the sonar hypothesis. If dolphin biology was disrupted by the sonar to the degree that these animals felt significantly disoriented, might they have consciously chosen to beach themselves? Terrible events may cause animals (including humans) to fall into such an acutely emotional state that they behave in ways that lead to their deaths. The time course involved may be brief or prolonged. Flint comes to mind here, the grieving young chimpanzee who died so soon after his mother. We have seen that any number of animals, from apes to rabbits, may respond to emotional trauma by shutting down emotionally.

With examples like these—the dolphin Kathy, the Cornwall dolphins, the chimpanzee Flint—we veer into the tricky area of animal mental health. To start with, not every example of self-harm, even in humans, is rooted in the urge to die. Sometimes, depression leads to the inability to care for oneself or to eat or sleep properly, but this situation may exist separately from suicidal wishes. There may indeed be no link at all between outright self-injury and suicide: the American Psychiatric Association notes that the unfortunate trend of cutting among adolescent girls, though a form of self-injury, is not a suicidal behavior. In fact, most mental health professionals see people who cut into their flesh with blades or knives as striving to help themselves (though in a dysfunctional and dangerous way that signals a need for help), as the stab of physical hurt temporarily relieves their deeper emotional pain.

Nor is self-harm limited to humans. We see it also in captive chimpanzees—and not only chimpanzees who are subjected to repeated biomedical procedures in labs. Scientists Lucy Birkett and Nicholas Newton-Fisher collected twelve hundred hours of data on forty socially housed chimpanzees in six zoos in the United States and the United Kingdom. While much of the apes' behavior was deemed normal, the abnormalities were sufficiently prevalent to be termed "endemic." Suicide was not reported, but chimpanzees rocked repetitively, bit themselves, plucked their own hair, and ate feces. Some of these behaviors occurred at low levels and for brief periods, but it's worth noting that every one of the forty zoo chimpanzees showed some sort of abnormality, while in 1,023 hours of focal animal sampling of wild chimpanzees in Uganda, not one of the abnormal behaviors was seen.

Given what we know of post-traumatic stress syndrome (PTSD) in wild elephants, however, I wouldn't rule out the possibility of abnormal behaviors in wild chimpanzee populations in regions where apes come under threat from humans. When elephants' early attachments within their families are disrupted because of poaching and war, the result is a breakdown of normal elephant behavior and culture. Gay Bradshaw and her coauthors (including long-term Amboseli elephant researchers) published a report to this effect in *Nature*. That elephants suffer from PTSD in elephant-killing zones emerges in part from their capacity to mourn their family members. With these elephants in mind, we can see that the zoo chimpanzees have turned their own wellspring of emotion back onto themselves: emotion felt in this case not for others' lives lost and bonds disrupted but for a life drastically limited in physical, cognitive, and emotional ways.

In this thicket of connections among animal depression, self-harm, and suicide, two entwined lessons stand out. First, our species is part of the problem and needs to be part of the solution. Compassionate response saved some of the stranded dolphins at Cornwall and undergirds activists' fight against the poaching of elephants for their ivory.

Compassionate awareness leads to the realization that many animals now held captive—elephants, great apes, and dolphins among them—should be living, if not in protected reserves in the wild, at least in sanctuaries. Even well-intentioned zoos simply cannot provide for the psychological health of these creatures. Bile farms that imprison bears go off the charts in the harm they do to animals; no such place should be allowed to exist at all.

The second lesson involves animal grief: We humans don't just study the phenomenon of animal grief. In a broad sense, we cause animal grief as well. We bring about conditions in the wild and captivity that lead animals to feel a sort of self-grief, and at times to feel empathy for others' suffering. Whatever caused that mother bear on the Chinese bile farm to run into a wall, in the end, it was human behavior—human greed twinned with an insensitivity to animal suffering—that murdered her.

12 APE GRIEF

It's November 22, 1968. Earlier that month, Richard Nixon had bested Hubert Humphrey in the US presidential election. A massive operation in Vietnam had initiated a sweep of the Ho Chi Minh Trail, resulting over time in the dropping of three million tons of bombs on Laos. Yale University began to admit women. And on this very day, the Beatles released their "White Album."

In the forests of Tanzania, the echoes of these political and cultural events are faint. Here, the air rings with chimpanzee pant-hoots. In the Gombe Stream population, chimpanzees, some excitable, others calmer, are on their way to becoming household names for animal-behavior aficionados: Flo, Fifi, David Graybeard, Goliath. These chimpanzees, already in 1968, have been observed for eight years by Jane Goodall—who is no longer dismissed as a *National Geographic* cover girl now that her discoveries of tool use and hunting have rocked the scientific world.

On this morning, Gombe researchers Geza Teleki and Ruth Davis follow a group of chimpanzees who walk through dense undergrowth. Serious students of ape behavior, Teleki and Davis are engaged to be married. Neither is aware, of course, that before the end of the year, Davis will be dead. Goodall makes an emotional acknowledgment to Davis and the "long arduous hours" she spent at Gombe in her book *In the Shadow of Man*. "It may have been due to physical exhaustion," Goodall writes, "that one day in 1968, Ruth fell from the edge of a precipice and was instantly killed. Her body was found only after a search of six days."

Davis was buried in Gombe. "Her grave is surrounded by the forest," Goodall notes, "and reverberates, from time to time, with the calling of the chimpanzees as they pass by."

What a terrible irony, then, that on this November morning, Teleki and Davis observe the immediate aftermath of a chimpanzee's death from a fall. They arrive at a clearing where chimpanzees, as Teleki would write in a journal article five years later, "explode into frenzied activity and raucous calling, including shrieks, cries, screams, waa barks and wraaah calls." In a dry stream bed within a gully, the chimpanzee Rix lies still. A necropsy will later determine that Rix's neck had snapped, causing instant death. Teleki and Davis seemingly just missed witnessing what must have been his dramatic plunge from a fig or palm tree, which probably occurred while he was eating or resting.

Teleki's article offers a blow-by-blow reconstruction of the events he and Davis observed between 8:38 a.m. and 12:16 p.m, which they recounted into a hand-held tape recorder for later transcription. Most striking in their account is the prolonged attention paid by sixteen chimpanzees to Rix's body—and how that attention varies. The chimpanzees are highly aroused in the wake of Rix's fall, but there's no uniform expression of that arousal. Just as we saw in Cote d'Ivoire, where the chimpanzee Brutus served as gatekeeper, determining which apes were allowed to approach the dead female Tina, the Gombe apes' personal ties and personality differences play a role in their responses to Rix's body.

Death has come suddenly to the Gombe community, and the apes' responses emerge as part of a fast-moving situation. In the little area around Rix's sprawled and lifeless body, a wild mix of behaviors is playing out. "Aggressive, submissive, and reassurance actions are performed," Teleki writes, "at high frequency and intensity by nearly everyone present, with many swift shifts in demeanor." Let's look at the males Hugo and Godi as a way of sampling some of the individual variation—and mercurial mood shifts—in the chimpanzees.

Hugo displays vigorously and, at one point, hurls several large stones toward the body, which do not hit Rix. Shortly thereafter, he stills himself (though his hair is still erected, a sign of arousal) and sits on a rock, where he's joined by another male. Hugo gets up, stands right next to the body, and stares at it for several minutes. Next he resumes his high-energy display, running away from the corpse. Later, in the area around the body, he mates with a female. When, much later, the male Hugh walks away from the death site, other chimpanzees, including Hugo, follow, after one last intensive bout of peering at Rix.

Godi, an adolescent, responds somewhat differently. He vocalizes more persistently than Hugo, uttering *wraaah* calls. Coming near the body, he stares at it while whimpering and uttering other vocalizations. To Teleki, he appears "extremely agitated, more so than any of the others." Throughout the next hours, Godi attends closely to the body. At 11:45, close to the time the group moves off, Godi is the only chimpanzee still watching Rix.

At first glance, it may seem that Hugo's and Godi's responses to Rix's death differed only slightly. Both showed signs of arousal, and neither one—in fact, no chimpanzee at the scene—touched the body at any time during Teleki and Davis's observations, toward the end of which the apes moved on. In this way, the Gombe chimpanzees, in their response to Rix's death, differ quite a bit from the Tai chimpanzees, in their response to Tina's death. At Tai, touch was an important element in the group's reaction.

Teleki makes a point of noting, though, Godi's "exceptional performance" that morning. Godi acted unlike the other chimpanzees in three ways: his proximity to the body, his level of agitation, and the frequency of his *wraaah* calling. *Wraaah* calls are "high-pitched, repetitious, plaintive wails," Teleki writes, "which can carry for a mile and more along the acoustic funnels of steep valleys, convey[ing] an intense emotional state which cannot be adequately communicated in words." Even though he didn't touch the body, Godi was emotionally affected by what had happened to Rix. Godi, notably, had often accompanied Rix in their daily travels.

It could also be argued that Godi's behavior derived from his heightened sensitivity to a kind of contagion that spread among the chimpanzees as they displayed, called, mated, and generally acted with high arousal around the body. Since *wraaah* calls are sometimes given when chimpanzees meet up with strange humans or with a Cape buffalo, or when two groups meet, as well as when dead baboons or chimpanzees are sighted, Godi's use of them doesn't really help us understand what he was feeling. Certainly we cannot make an airtight case that Godi or any of the other apes recognized Rix's death for what it was. "It remains uncertain," Teleki concludes, "whether any participant grasped the conceptual difference between life and death."

Yet insistent questions intrude, based on the cumulative weight of what is known about chimpanzees and death. Why wouldn't a social partner of the deceased ape feel strong emotion at the sight of his confederate lying lifeless? Why wouldn't a community made up of intensely social beings respond as a community when one of its own dies? The choreography of tight ties within families and between social allies is the ever-constant backdrop of events in chimpanzees' lives. We cannot understand what a chimpanzee does apart from the social dynamics that surround him any more than we can grasp a person's behavior by looking at him in isolation. And we have seen what happened when Tina died at Tai. Tina's little brother, Tarzan, clearly felt some emotion, and he was allowed to express it precisely because the male Brutus, who took charge of which chimpanzees were allowed near the body, recognized Tarzan's kinship status vis-à-vis Tina.

Nudged by such extraordinary, all too rare observations of chimpanzee responses to death in the wild, zoo scientists are paying keen attention to deaths in captivity. At a Scottish safari park, when an old female chimpanzee fell ill, zoo staff anticipated her death and switched on the videocamera. At the park, two mother-offspring pairs lived together: Pansy, the dying female, whose age was estimated to be in the fifties, and her daughter Rosie, age twenty, plus Blossom, a female about Pansy's age, and her son Chippy, age thirty. The apes were in their heated winter quarters when Pansy, who had been listless for some weeks, began to breathe in a labored manner. Pansy's companions seemed aware that something was amiss; in the ten minutes before her death, they groomed or caressed her at what the observers judged to be a higher than usual rate. Right around the presumed moment of death, group members continued to be highly active. In the journal *Current Biology*, James Anderson and his colleagues wrote up what happened with admirable precision:

16:24:21 Chippy crouches over Pansy's head then appears to try to open her mouth. Rosie moves towards Pansy's head.

16:24:25 Blossom, Chippy and Rosie simultaneously turn toward Pansy's head. Chippy and Rosie are crouched over Pansy's head. Chippy pulls Blossom's face down toward Pansy's.

16:24:36 Rosie moves from Pansy's head toward her torso. Blossom moves away from Pansy. Chippy lifts and shakes Pansy's left shoulder and arm.

The chimpanzees continue to caress and groom Pansy. At 16:36:56, Chippy "jumps into the air, brings both hands down and pounds Pansy's torso, then runs across and off the platform." This startling behavior departs significantly from the male Hugo's rock-throwing displays around Rix's body in the wild; here, Chippy attacks Pansy directly. Yet at Tina's body, some males also displayed aggressively around the corpse and even dragged it short distances. When the male Ulysse moved Tina's body about two meters, it was Brutus who dragged it back to its original location.

Chippy's behavior, then, isn't far outside the range of what male chimpanzees do in the wild when a death occurs. Was he expressing anger or upset? Was he trying to elicit some kind of response from his immobile cagemate? Both possibilities seemed plausible to Anderson and his coauthors.

The behavior of Pansy's companions remained atypical through the night and beyond. (Data of this sort are unlikely to come from observations made in the wild, because wild chimpanzees fairly soon move away from the corpse.) Pansy's survivors slept fitfully. Her daughter, Rosie, stayed near her body. In contrast to the moments before her death, no one groomed Pansy's body, although Chippy attacked her corpse three more times during the night.

The next day, the survivors were "profoundly subdued." In silence, they watched as zookeepers removed Pansy's body. For the next five nights, none of the chimpanzees slept on the platform where Pansy had died, even though they had favored that spot in the past. Indeed, for weeks they remained quiet and ate less than usual. These signs of animal grief—the altered routine, the disturbed mood—are by now familiar.

Observations like these matter not just for our understanding of apes but for apes' lives themselves—that is, for the dignity with which we treat them even as we keep them captive. The growing data bank of primate responses to death is fomenting a revolution in how captive primates are treated when one of their group dies. As we saw at the Scottish park, apes may be offered a chance to spend time with a deceased companion, and to watch when later the body is taken away.

At Chicago's Brookfield Zoo, the gorilla female Babs suffered from an incurable kidney disease and was euthanized at age thirty. The staff there organized what they called "a wake" for Babs's companions. Gorillas from different generations were present, some of whom were visibly emotional. The Associated Press reported the event in this way:

> Babs' 9-year-old daughter, Bana, was the first to approach the body, followed by Babs' mother, Alpha, 43. Bana sat down, held Babs' hand and stroked her mother's stomach. Then she sat down and laid her head on Babs' arm. . . . Bana rose up and moved to Babs' other side, tucked her head under the other arm, and stroked Babs' stomach.

We recognize here the grief of a child confronted with the utterly still body of a loved parent. For the whole of Bana's life, she had been near her mother. Her need for touch, to feel her mother in a literal sense, is notable: we primates are tactile creatures. Babs's other cagemates approached her too. Nine-year-old Koola brought close her infant daughter, a baby who had received Babs's affection during her young life.

Ramar, a silverback male, age thirty-six, stayed away from Babs. Some gorilla males are aloof in this way; others are not. At Boston's Franklin Park Zoo, the gorilla female Bebe was euthanized to spare her the pain that accompanied the malignant masses in her body. Diane Fernandes, then curator of research at Franklin Park and now director of the Buffalo Zoo, remembers the response of her mate:

> We first let the male Bobby in with the body and he did try to revive her, touching her gently, vocalizing and even placing her favorite food (celery) in her hand. When he realized she was dead, he began to call in this soft hoot but then started to wail and bang on the bars. It was clearly a demonstration of immense grief and it was very sad to watch.

Fernandes shows no hesitation in deploying cognitive and emotional terms for what Bobby experienced when Bebe died. She says that he realized the fact of her death. As opaque as Bobby's thought processes must remain to us, the sequence of his behaviors supports this conclusion. The gift of Bebe's favorite food was probably offered because Bobby thought, or hoped, that Bebe was alive, or because he wanted to somehow encourage her, to revive her with a sensory experience she loved. When this strategy failed to work, Bobby erupted into sadness.

The Franklin Park Zoo allowed Bobby some time alone with Bebe's body, and then three other gorillas were permitted to approach. These apes also, Fernandes recalls, touched Bebe's body, "as if to rouse someone who was sleeping." But unlike Bobby, they did not vocalize. Perhaps they didn't make the cognitive leap that I suspect Bobby did, or perhaps they just expressed their grief differently.

From the Franklin Park event emerge some questions that might guide future research: Do survivor apes routinely try to revive the dead one? Do those attempts then cease because, as apparently happened with Bobby, a cognitive leap is made and the fact of death is comprehended? Or do the survivors instead continue to search for the ape that has died? Do mourning behavior and searching behavior ever coexist in the same animal? And how do these variable behaviors play out across individuals with different relationships to the deceased?

Having logged hundreds of hours observing, filming, and analyzing behavior in gorilla families, I am not surprised that zoo staff "believe in" gorilla grief even though many questions remain unanswered. As Pittsburgh Zoo keeper Roseann Giambro puts it, she knows in her heart, because of what she's seen, that gorillas mourn. I understand this sentiment, and at the same time know it must become the basis for hypothesis-testing. Zoo staff could record not just the actions taken by the gorillas but the quality of the actions: the heaviness in the muscles of a grieving ape, the anxiety in the movements of an ape searching for another who is missing, the frantic or mournful tone in a call shared with the group (or of course, the absence of these attributes). Often overworked, zoo staff may be hard-pressed to supplement their daily responsibilities with the task of compiling a detailed timeline of gorillas' responses to death and notes on the quality of those responses. Yet that's the best way to find out if the gorilla Bobby's two-stage reaction—first trying to revive the dead animal and then grieving—might be repeated elsewhere when captive apes die.

At Pittsburgh, two deaths, eight years apart, are fixed in keeper Giambro's mind. In 1997 a female called Becky died. The causes remain undetermined, but Becky was in her mid-forties, which represents a good life span for a gorilla. For weeks afterward, her closest companion, Mimbo, stared at the spot where she had died and refused to walk through that room. Tufani, a younger female, responded differently. In an anxious state, screaming and rushing around the enclosure, she searched for Becky. It seems likely to me that Mimbo, with the wisdom of age and experience, grasped that his friend was gone, as Bobby apparently did with Bebe. Tufani, by contrast, may have lacked such understanding because she was younger and death was newer to her. Alternatively, the two gorillas' responses may point us once again to variation in mourning habits according to personality.

The silverback male Mimbo also lived to his mid-forties. When he died of liver problems, his son Mrithi, age thirteen, pushed at the body with his hands and feet. The female Zakula, who had borne three offspring by Mimbo, also pushed at his body, as if urging Mimbo to rise, and she groomed him. The gorillas vocalized in a way that was unusual and that sounded "mournful" to Giambro. Eventually the gorillas went outdoors, and zoo staff removed Mimbo's body. When the apes reentered their enclosure, they searched for the silverback. For a week, group members were significantly disrupted in their behaviors, including their eating patterns. Gradually, Mrithi was allowed by the females to assume a position of leadership, and the gorillas' lives settled down.

After Mimbo's death, there was no clear shift—as I think probably happened with Bobby after Bebe's death and with Mimbo himself after Becky's death—from a sort of "search and rescue" mode to a mourning mode. Mimbo's groupmates only gradually came to accept his absence, despite having viewed his body directly after he passed away.

Even at this early stage, with so much more to discover about apes' responses to death in captive situations, one big message comes through. An ape may die gradually, becoming weaker and weaker with illness or age. Or an ape may die suddenly—because his heart stops, or because of an unexpected outcome on the surgical table, or because loving human caretakers move quickly to end pain once death is inevitable. The survivors deserve a chance to sit with the body, and to touch the body if they so wish. The outcome of this process will differ according to the kinship status and personality of the one who has died, and almost certainly with the age and knowledge level of the survivors as well. But whatever the outcome, the offered opportunity is a kindness deserved by primates who form tight ties and mourn their losses.

When Rix died in the Gombe woodlands, the chimpanzees' response to his sudden stillness was observed with acute sensitivity by Geza Teleki and Ruth Davis. Soon after, Davis suffered her own fatal fall. I do not know Teleki personally, but I wonder if, back in 1968, when his own grief was fresh and terrible, he felt a connection with these apes he knew so well, who had recently confronted sudden death themselves? I believe that Teleki would find such a question respectful to the memory of his fiancée. We grieve as primates, and we have company.

13 ON BISON DEATH IN YELLOWSTONE AND OBITUARIES OF ANIMALS

Paradoxes thrive in Yellowstone National Park, a wilderness area flung across vast tracts of Wyoming and bits of Montana and Idaho. At the park, I walk inside the caldera of the world's most explosive volcano and marvel at its power. When the Yellowstone volcano erupts again, as geologists tell us it will (it's already overdue), the ash will alter the earth so thoroughly that most living creatures won't survive. Even now, the earth growls and spits at Yellowstone, its controlled power on display.

At the same time, the park seethes with life: bison, bears, elk, moose, coyotes, wolves, and birds allow a visitor to see a dynamic ecosystem in action. In spring and summer, the valleys and uplands bound with infants. Those of the mammalian variety rush about on ungainly legs, then seek their mothers' milk. The cute bison and elk babies photographed in one season stand a decent chance of becoming wolf or coyote meals in the next. No managed parkland or tame zoo, Yellowstone is a place of life-and-death struggle. On a smaller scale, this struggle plays out in our backyards too—ask anyone with a domestic cat who returns home with "gifts" of half-consumed birds, moles, or frogs. The 2.2-million-acre scale of the place, though, and the glorious faunal diversity in it, make Yellowstone an incomparably intriguing place for any nature lover.

Visitors' senses kick into high alert at Yellowstone for another reason too: even aside from the volcano, it's a dangerous place for our kind. *Death in Yellowstone: Accidents and Foolhardiness in the First National Park* by Lee H. Whittlesey is an oddly mesmerizing chronicle of the many ways one can perish in a gorgeous setting. Dotting the Yellowstone landscape are pools of brilliant color, sapphire blue and sunshine yellow. In them live extremophiles, tiny microorganisms that thrive on intense heat and that may cause an extreme death for the unwary or the impulsive.

In 1981 two twentysomething California men visited the park's Fountain Paint Pot area. One of the men's dogs leaped off the walking path into a 202-degree hot spring called the Celestine Pool. What began as a canine tragedy became a human one when one of the men, ignoring bystanders' pleas, dived right into the near-boiling water. The man emerged without the dog. Already by that point, it was too late for either of them. The dog died inside the pool. The man staggered about, with eyes totally white (and blind). Another visitor tried to help by removing the man's shoes, but the skin peeled off along with the shoes. Later, "near the spring," Whittlesey reports, "rangers found two large pieces of skin shaped like human hands." First taken to the clinic at Old Faithful, the man was soon transported to a hospital at Salt Lake City. He died there the following morning.

At the end of the size range opposite the hot-spring extremophiles are the iconic National Park animals, the American bison. Some people, Whittlesey notes, see the buffalo less as a potentially dangerous animal than as a romantic symbol of a vanished American past. "Many visitors," he writes, "want to approach it, to touch it, to somehow establish a close link with it, as if that might somehow connect them to their own frontier heritage." Unfortunately, a close approach to a buffalo is most likely to connect a person with a piercing set of horns. The Park Service warns people about this reality via posted signs and distributed brochures, but romanticism sometimes trumps common sense. Yellowstone's first recorded death-by-bison occurred on July 12, 1971. A thirty-year-old visitor from Washington state yearned to photograph a solitary bull lying in a meadow, and approached within twenty feet of the animal to do it. The bison charged, tossing the man more than twelve feet with a powerful flip of the horns, which tore apart the man's abdomen and injured his liver. His death unfolded in front of his wife and children, even as the family had in its possession a red "danger" pamphlet warning of too-close encounters with the park's wild animals. Nowadays, maybe a print brochure isn't enough; a YouTube video of bison-tossed tourists, sent to visitors' cellphones, might do the trick more effectively.

But it's not just the foolhardy who are at risk in Yellowstone. During my visit to the Park in the late summer of 2011, all the talk was of grizzly bears. In a twelve-month period, three hikers had been mauled to death by bears who make their home in Yellowstone and who responded to a human's presence, in what is essentially their living room, as a bear might be expected to respond.

The joy of Yellowstone, though, is that it invites a shift in perspective, away from ourselves and toward other creatures. Do any Yellowstone animals grieve when one of their numbers dies? Reading a travel article in the *New York Times*, I learned that bison sometimes fall into the boiling hot springs. Shimmering up from the hot depths, their bones tell a story of sudden, accidental death. Do other bison ever witness these scalding deaths and turn away in sorrow? We don't know. But the bison, I think, hold a key to asking good questions about mourning patterns in Yellowstone.

On my first visit to Yellowstone in 2007, I fell hard for the bison, an animal of keen importance to humans for many millennia. Cave painters rendered bison in true-to-life images throughout Ice Age Europe, showing the acute perceptual powers of our ancestors in regard to the natural world. But bison weren't represented only in realistic ways. A startling image in France's Chauvet Cave, created by an artist living thirty thousand years ago, depicts a creature that is half buffalo and half human woman. Our hunting-and-gathering ancestors thought symbolically with animals in ways that remain beyond our ken but continue to stir our imaginations.

Observing the Yellowstone bison brought me back to my Kenya days. Following the Amboseli baboons on foot to collect data on their feeding patterns, I often encountered elephant, lion, leopard, hyena, warthog, rhino, and even the occasional Cape buffalo. Back then, I did everything I could to avoid proximity to the African buffalo (not to mention the big cats). I was a vulnerable biped on the savannah, in awe of massive horned beasts who could gore me or worse. But in Yellowstone, where I could ride in a vehicle—and deploy common sense when stepping out of it—I couldn't take my eyes off the American version, the bison of the Great Plains.

Our usual procedure when bison-watching is to drive out on Yellowstone roads to Hayden or Lamar Valley, sight a bison herd, and pull over to the side of the road. The bulls are shaggy, snorting, and solidly built. The females and babies, lighter of foot, move together in the timeless mammalian dance of suckle-and-wean: long after the mothers are ready to nudge them toward independence, the babies want to keep nursing. Bound by an invisible cord, these moms and youngsters remind me of baboons: a baby cavorts away from her mom, twists and jumps in play, then seems to suddenly realize she is out of her comfort zone, and zooms back to home base.

It is a thrill to observe the large Yellowstone herds. After the terrible toll inflicted by the slaughter of the late nineteenth century, only twenty-five bison survived—the sum total in the entire United States, and all located in Yellowstone. The Yellowstone Buffalo Preservation Act, introduced to Congress in 2005 but never passed into law, noted that those survivors' offspring today "comprise the Yellowstone buffalo herd and are the only wild, free-roaming American buffalo to continuously occupy their native habitat in the United States." Compared to ranched buffalo, whose genes have long been intermixed with those of domestic cattle, the Yellowstone buffalo are genetically unique: pure and wild.

Do these buffalo, killed in terrible numbers and in terrible ways by humans, respond with emotion to the natural deaths of their own kind—death by disease, by predator, by old age, by a stumble into a hot pool? Biologist John Marzluff, whose work on corvids I discussed in chapter 8, brings a glimmer of light to this subject. With a group of his students, Marzluff surveyed a recent wolf kill site at Yellowstone. Over the previous two weeks, terrestrial and avian predators had reduced to bone the carcass of an old female bison. Near the skeletal remains was a boulder, "split in half," Marzluff writes, "by eons of freezing and thawing."

As teacher and students stood at the site, in thundered a bison herd, heading right for the carcass. Not lacking common sense, the biology group retreated and watched from a distance. The bison stayed for nearly an hour. "Each of the three dozen animals walked up to their former companion's bones and smelled them," Marzluff reports. "They sniffed the remains and the soiled snow and dirt." Departing, they made their way right through the split boulder and walked out of view. "These animals," Marzluff concludes, "are still sensitive to a past event." Marzluff's account may remind us of the African elephants who caress the bones of loved ones. Given the emotional power entailed in such events we can grasp why Marzluff describes what he and his students saw as "sacrosanct."

It's our old refrain by now, but true still—there's been little science done so far on bison grief. One segment of *Radioactive Wolves*, a television documentary about the thriving of wildlife in the wake of the 1986 Chernobyl nuclear disaster, focuses on a pack of wolves approaching the remains of a bison calf. The wolves didn't kill the calf—it was already dead on the ground. Scavengers as well as hunters, the wolves begin to tear into the small carcass. The bison regroup then, and chase the wolves off. I snapped to attention when the narrator stated that the adults were "mourning" the calf, as is "typical" for bison.

But where is the science to support this notion? One classic text on this animal is *American Bison* by Dale F. Lott. Nowhere in its thorough index do the terms "death," "grief," or "mourning" appear. The whole arena of animal emotion is tough enough to study in the wild, and the old bugaboo of anthropomorphism still prevents some scientists from even trying to collect the needed data. Look at what Lott calls the first section of his book, though: "Relationships, Relationships." Bison are herd animals, living in social conditions ripe for the formation of strong social bonds and for mourning. Writing about the high drama of the breeding season, Lott has this to say: "Attraction, rejection, acceptance, competition, and cooperation within and between the sexes create vital, compelling, generally short-lived, and shifting relationships." Short-lived: we aren't talking about long-term monogamous bonds here. But male-female bonds do exist, and of course mother-infant bonds do too.

An interviewer once asked me what I'd do if I had unlimited funds to study animal grief. The short answer? I'd make my way to Yellowstone with those funds and a massive cache of patience. Death doesn't happen in bison or other animal groups in front of a casual observer's eyes; to be present at the right moment, or soon after, would take good fortune as well as persistence. But as we have seen, clues to bison mourning are already available. Moose behavior tantalizes too. Biologist Joel Berger has labored in some of the world's most unforgiving (and cold) landscapes, from Yellowstone to the Russian Far East and Mongolia, in order to learn about animal behavior. In Yellowstone, he focused in part on moose. "Just as parents know the behavior of their children," he writes in his book *The Better to Eat You With: Fear in the Animal World*, "my intent was to understand each moose."

One frightened moose orphan ran for more than half a mile when Berger wanted to put a radio collar on her, then halted at the precise spot where her mother had died. Another moose, this one a mother, returned repeatedly to the spot where her calf had been struck by a car, "apparently searching for her missing calf." What would happen if Berger, or other scientists like him, staked out the carcass of a moose who had died of natural causes, then watched over days and weeks what other moose did as the body gradually turned to bone? Would the moose detour to view the bones of a lost herd member, as Marzloff's bison did and (see chapter 5) as elephants do? When bison, or moose or other animals, encounter bones of their own kind, do they inspect the bones in a detached manner, or do they feel something as they look? Could we humans tell the difference? Would an experienced observer of bison note clues to an emotional response in bison after a death in the group?

Do animals read bones on the ground like we read obituaries? Is that too fanciful a thought? I'm not so sure that it is. Well-written obituaries elegantly compress a life into a few descriptive passages. In a parallel but nonlinguistic way, the bones an animal leaves behind may do the same. The compression of obituaries may bring sadness or even a hint of futility to readers. Can eighty long years of a life really be reduced to eight short paragraphs? Vital force can be unleashed from these life haikus, though, as I have discovered by reading regularly the *New York Times* obituaries section. This habit might be thought elitist in that the *Times* skews its death reporting toward the famous, yet for me it affords a chance to learn about fascinating lives I otherwise wouldn't encounter.

When a woman named Martha Mason died at age seventy-one, she had dwelled for six decades in an iron lung. Paralysis had struck Mason as a child, a result of polio; from that point on, her habitat was, as the *Times* described it, a "horizontal world, a 7-foot-long, 800-pound cylinder." In the obituary photograph, Mason's white-haired, bespectacled head juts from one end of the machine, which is lined with porthole windows and resembles some sort of deep-sea research craft. And from it, Mason did explore the world. While encased in the iron lung, Mason attended Wake Forest College, hosted dinner parties, and employed, in later years, a voice-activated computer to write *Breath*, her memoir.

When my impatience meter ratchets up in the face of some trivial annoyance, I sometimes think of Mason. Faced with anything but trivial challenges, she did more than just endure; she lived with courage and verve. In this way, reading obituaries can inspire me. It's no surprise that I'm moved especially by the lives of people who loved animals. In Vermont, the artist Stephen Huneck built the Dog Chapel, where people and their dogs may seek moments of interspecies serenity. The church's windows feature stained glass and dog images; blanketing the walls are handwritten notes of grief, describing pets sorely missed. Atop the steeple sits a winged Labrador. I only wish Huneck had found serenity of his own within the chapel. In despair at having been forced to lay off most of his art-business employees, Huneck committed suicide at age sixty-one.

At that same age, in 2012, Lawrence Anthony died of a heart attack. In 2003, shortly after the US invasion of Iraq, Anthony had saved the lives of thirty-five starving animals in the Baghdad Zoo. Of 650 animals who resided there at the start of the war, these were the sole survivors. Anthony also restored the zoo itself to decent condition. I had known a little about Anthony's work before, but his obituary colored in the broad outlines of a famous life. Working for animal conservation in Africa, Anthony rocked out to Led Zeppelin and Deep Purple as he crisscrossed the countryside in his Land Rover. He forged a deeper connection with elephants than with any other animal, and his obituary concludes on a mystical note: "The elephants also survive him. Since his death, his son Dylan told reporters, the herd has come to his house on the edge of their reserve every night."

Like a hall of mirrors, an obituary illuminates in our imagination not only the person now lost, but also a life that echoes (and echoes again) across time and space. In it we read the names of those who died before and of the survivors, the past and future in unbroken continuity. *Dandelion Wine*, Ray Bradbury's masterpiece portrait of one boy's summer in 1928 small-town Illinois, captures this life-within-death theme. One hot night, twelve-year-old Douglas Spaulding comes to grasp death's inevitability when he releases a jarful of fireflies to evanescent freedom. "Douglas watched them go," Bradbury writes. "They departed like the pale fragments of a final twilight in the history of a dying world. They went like the few remaining shreds of warm hope from his hand."

Douglas's dying grandmother imparts a lesson to him that resonates with us today. Douglas sits on her bed in the family home. He cries, knowing that soon she will leave him forever. She tells him:

> Important thing is not the me that's lying here, but the me that's sitting on the edge of the bed looking back at me, and the me that's downstairs cooking supper, or out in the garage under the car, or in the library reading. All the new parts, they count. I'm not really dying today. No person ever died that had a family. I'll be around a long time. A thousand years from now a whole township of my offspring will be biting sour apples in the gumwood shade.

There's a phrase to arrest the heart: *No person ever died that had a family*. It is as apt for an animal who dies without an obituary as for a person who dies with one. People often commemorate an animal's loss by coming together in some type of symbolic ritual, as we saw in chapter 10 with the German polar bear Knut, on a national scale, and the cat Tinky, on a family-and-friends scale. We fix memories of these special creatures in our minds, and pass them along to others in our generation and the next.

And sometimes, animal *are* memorialized in obituaries. The only obituary I have yet written was for an ape. When the chimpanzee Washoe died in 2007 at age forty-two, the American Anthropological Association (AAA) asked me to compose a notice for its monthly newsletter. Based on her groundbreaking accomplishments learning signs and phrases in American Sign Language, Washoe was judged by the AAA to be a figure of importance to its members. Impressed by the slightly transgressive nature of this request, and agreeing with the AAA's assessment of Washoe's stature, I wrote the obituary.

Wild-caught in West Africa, Washoe as a youngster was brought to the United States. She eventually came to live with psychologists, first Beatrix and Allen Gardner and then Roger Fouts, at a number of academic institutions including the University of Oklahoma, where, as a graduate student, I met her. Washoe overturned species-bound assumptions about who can deploy language to communicate and who can't. Growing up immersed in human culture, Washoe learned a modified American Sign Language. She signed creatively, as when she coined the phrase "open food drink" for "refrigerator," and she molded the hands of her adopted son Loulis so that he too would learn signs. Going well beyond the expression of simple desire for favorite foods, Washoe conversed with those around her, as when she expressed empathy for Fouts, her closest human friend, when he had broken his arm.

The page layout for the January 2008 issue of *Anthropology News* that contains Washoe's obituary is a study in boundary maintenance. Across two facing pages in the "Rites of Passage" section are death notices for five accomplished anthropologists, aged fifty-seven to ninety-four. Turn the page and there, all on its own above the "Kudos" section that congratulates AAA members for honors received, sits my little article. "Also Noted," the headline reads: "Washoe, age 42." In this way, a nonhuman animal is included with anthropological luminaries, but by physical placement and subtle use of language she is, at the same time, kept apart. This editorial decision I can understand. Were you the spouse, or child, of an anthropologist who had recently died, would you welcome the sight of his life story and photograph pushed right up against Washoe's, who appears with her simian face and robust brow ridge? (Well yes, I would, but that's likely a minority view.)

Constrained by space, I failed to include in the obituary any mention of Washoe's survivors, notably her son Loulis. Yet I did address her legacy, the past-to-future continuity of which the fictional Douglas Spaulding's grandmother spoke, albeit in another way:

> As with a human, it is impossible to sum up Washoe's life by reference to academic debates and publications. Her personality (and her well-known interest in shoes and shoe catalogs!) made her a unique individual. Messages from Australia, Belgium, Italy, Mexico and elsewhere, posted on a memorial page for Washoe, reveal her impact on persons around the globe. Reading these tributes, one gets the sense that Washoe's enduring legacy comes not from the number of signs she could be said to ac-

> quire, or whether those signs amounted to language. Rather, it relates to how she caused people to think hard about the dividing line between apes and people, indeed, about the very notion of ape personhood.

Animals like Washoe who live in the public eye may catalyze shifts in our thinking about what makes a human being, but not an ape or dolphin, deserving of the term "person." Flo, arguably the most famous wild chimpanzee in history, had the same effect. Through Jane Goodall's early dispatches from Tanzania, Flo's maternal skill and inexhaustible patience with her babies and her juvenile son Flint captured the public's imagination. When Flo died in 1972, her obituary appeared in the *Times* of London.

When animal celebrities die, few people seem to object to a newspaper's stretching its "obituary" category to include them. When it comes to obituaries for our pets or other companion animals, the response may be quite different. Anthropologist Jane Desmond has written about the power of such obituaries to subvert the animal-human boundary and thus to unnerve a healthy segment of the human population. Some years ago in the *Iowa City Press-Citizen*, Desmond's local newspaper at the time, an obituary was printed for a black Labrador named Bear—the first animal obituary published by the paper. Bear, who frequently walked along, and napped on, the town streets, had been known to many people. Even so, that brief obituary, writes Desmond, "became the cause of bitter debate" in the community. Especially offended was a woman named Sue Dayton, whose sister-in-law's obituary had appeared on the same page as Bear's. Discord erupted in the town as words like "distasteful" and "disrespectful" were hurled around to describe Bear's printed memorial.

Desmond considers why a published obituary should incite such negative emotions when other customs that honor deceased pets do not. Through physical or virtual pet cemeteries, or online memorial pages for beloved animals, people cordially share their pet memorials with others of a similar mind. Newspaper obituaries, by contrast, are a highly visible matter of public record. "They do not have to be sought out specifically," Desmond writes, "but rather land on our table, in the news pages flopping open by the bacon and eggs, inserting themselves into every household." Because they openly announce that a pet was part of a family, and bring legitimacy to mourning a pet as a family member, obituaries for animals push up against the definition of "family" in ways that may be quite upsetting for some people. Writing for the *St. Louis Post-Dispatch*, columnist Betty Cuniberti lamented the pet-obituary practice by imagining "a sorrowful son opening our newspaper to look for his mother's obituary and finding her picture next to one of a hamster." Like Desmond, I think that Cuniberti's choice of a hamster was calculated to mock the idea of pet obituaries.

The pet obituary, then, upsets some of us while comforting others. I'm inclined to find comfort in any animal obituary. It's the animals themselves, not their obituaries, who trample an assumed animal-human boundary. This is as true with grief behavior as it is with cognitive accomplishments such as tool use or cooperative problem-solving. We know this from bereaved monkeys with their strong physiological response to loss, from the cat who wails in grief for her lost sister, from the horses who circle the grave of a deceased friend, from the buffalo who diverted their course to be with the bones of a lost female, and from the elephants who turn loved ones' bones over and over in their trunks. Desmond gets it right when she says, "As with humans, pet obituaries assign value to a life, define its highlights, extol socially validated accomplishments, and serve as models of living."

Assign value to a life. The language of the obituary is not the language of other animals. But doesn't that phrase capture precisely what animals do when they grieve? They assign value to a life once lived, a life now mourned.

14 WRITING GRIEF

This is shock to me—that the unremitting cold of the season of Ray's death—New Jersey sky like a pot carelessly scoured, twilight easing up out of the drab earth by late afternoon—is yielding by slow degrees to spring.

The widow doesn't want change. The widow wants the world—time—to have ended.

As the widow's life—she is certain—has ended.

JOYCE CAROL OATES, *A Widow's Story*

Inside me [some weeks after my wife Aura's death], lodged between spine and sternum, I felt a hard hollow rectangle filled with tepid blank air. An empty rectangle with sides of slate or lead, that's how I visualized it, holding dead air, like the unstirred air inside an elevator shaft in a long-abandoned building. I thought I understood what it was, and told myself, The people who feel this way all the time are the ones who commit suicide.

FRANCISCO GOLDMAN, *Say Her Name*

Memoirs of mourning have exploded, these last few years, into the white light of publishing fame. These are no third-person scholarly tomes about people's responses to death across time or among different cultures. Books like that, with their restrained prose and orderly footnotes, can be found on the shelves of anthropologists, psychologists, sociologists, and historians.

I refer to a different genre altogether: those shatteringly personal, I-mourn-my-loved-one-here-before-your-eyes memoirs, books that penetrate into our hearts because we know that, at some point in our lives, we too will become experts on their subject in the way we dread most. (As a writer, I'm choosing to focus on literary grief. In the third chapter of *The Nature of Grief*, John Archer expands this focus to review grief in film, the visual arts, and music as well as a different set of literary works.)

When grief slams into a life, the background hum of daily routine vanishes. "Grief has no distance," writes Joan Didion in *The Year of Magical Thinking*, her memoir of the year following the sudden death of her husband, John Dunne. "Grief comes in waves, paroxysms, sudden apprehensions that weaken the knees and blind the eyes and obliterate the dailiness of life." Writers, people who have spent their lives making meaning through the flow of words on the page, recover some of that dailiness by binding some of their grief to paper.

The genre itself is hardly new. C. S. Lewis wrote *A Grief Observed* in 1961. By that time, Lewis was "the most popular spokesman for Christianity in the English-speaking world," as one documentary noted. For decades, he had lived a don's intellectual life, and the life of a bachelor. Then Joy Davidman Gresham, an American poet and novelist, wrote to him from across the sea. Probing her own atheism, indeed beginning to leave behind her atheism, Gresham was drawn to Lewis's Christian perspective. Eventually Gresham and Lewis met. Relating to each other only cerebrally at first, they eventually fell in love. Lewis's intellectual equal—for that is what he considered Gresham—now brought him the very emotion for which she was named.

By the time the two married in 1956, Joy's cancer diagnosis had already intruded upon their lives. Her death came only four years later. *A Grief Observed* was published the following year, under the pseudonym N. W. Clerk. In the text, Joy is referred to as "H." (her legal name was Helen). Later, the book was reissued under Lewis's name, and by then everyone knew who "H." really was. Lewis's initial circumspection, his strong desire for privacy as he shared his wildest emotions and unwonted doubts, is a point I will revisit in a moment.

Based on jottings in four notebooks that he kept after Joy's death, *A Grief Observed* showcases a brilliant mind at once both blunted and sharpened by grief. A powerful cry arises from the first few pages. It's not a loss of faith in God that's the problem, Lewis writes, it's the revelation that he now believes "dreadful things" about God. He anguishes too over what he considers to be the inevitable dimming in his mind of the real Joy: "Already, less than a month after her death, I can feel the slow, insidious beginning of a process that will make the H. I think of into a more and more imaginary woman."

It's here, I think, that the human experience of grief begins to depart from that of other animals. Joy's death plunges Lewis into new anxieties, and a broad reassessment of what he thought he knew and believed. In grief's grasp, he relentlessly revisits the past and anticipates the future. He wrestles with questions that have no answer. Interestingly, in this context, he also remarks about "that terrible oxymoron, a 'spiritual animal.'" Lewis assumes that our species alone is capable of self-transcendence and awe in the face of the unknowable. He may be right, but I don't wish to assume that no other self-aware animals experience a glimmer of spiritual feeling. Jane Goodall, for one, famously thinks that chimpanzees may have their spiritual moments, based on their behavior at rushing waterfalls. Indeed, she goes further than I would in suggesting that chimpanzees as are spiritual as humans but lack a way to analyze or describe their awe and wonder. Chimpanzees' rock-hurling, vine-swinging displays at waterfalls ("the rain dance") impress me less in this regard than do their moments of quiet reflection, when their eyes track the falling water and they seem lost in thought.

Goodall's thoughts aside, it's clear that Lewis, or any one of us humans, grapples with grief in ways fundamentally different from the ways other animals do. In drawing such a stark dichotomy, I may seem to break with the tenor of the stories in this book. Yet as I noted in the prologue, to acknowledge that we humans think and feel differently from other living creatures need not amount to a manifesto of human superiority. To any such dismissive claim, the stories collected here shout a decisive "No!" We humans aren't superior to other animals because we grieve differently, any more than a self-aware animal like a dolphin is superior to an animal, like a goat, who is less able to reflect upon her own life.

Why shouldn't our grief be different? Evolutionary theory predicts species-specific behaviors in each animal. We humans don't erupt into displays of aggression around dead bodies as chimpanzees may; chimpanzees don't tell each other stories about the dead. Oh yes, chimpanzees may communicate with each other about a death in some way—we're only beginning to ask those questions. But they aren't the storytellers that we are, passing down elaborate narratives about our grandparents and parents to our children and grandchildren. Does that mean our grief is deeper than the grief of chimpanzees? Questions like this one miss the point. We each are what we are, animals bound together by our various ways of grieving.

Some self-aware animals, including great apes, elephants, and cetaceans, do remember past events and plan for future ones. Perhaps when individuals of these species mourn, they replay in their minds memories of time spent with the loved one. If so, those memories may not take on the vivid specificity that ours do, primed and sustained by the language in our heads: the sun-ripened image of a picnic in the forest, or the skin-on-skin feel of snuggling together on a cool morning. As the writer Temple Grandin argues, other animals' thoughts may be visual and impressionistic, less precise than ours about time and place and more invested in the cocoon of feeling that memory brings on. Do animals dwell on their sadness, closing their eyes at night aware that the blanket of grief will still be there at dawn? The answer is probably no. A Sisyphean sense that grief will be our partner, today, and tomorrow requires a faculty of self-examination that is beyond the ability of any species but our own.

The terrible power of this kind of self-knowledge can be found in Lewis's *A Grief Observed*. "I know that the thing I want is exactly the thing I can never get," Lewis writes. "The old life, the old jokes, the drinks, the arguments, the lovemaking, the tiny, heartbreaking commonplace." I get the sense, though, that Lewis desired less to report the contours of his grief than to tunnel deeper inside himself by writing about it. Recall that Lewis hid his identity when he first wrote the book. In this way, his book stands apart from many memoirs in the contemporary grief genre. Lewis didn't set out to make a public, wild lament, and his grief touches me more deeply as a result.

Lewis says something very interesting near the end of *A Grief Observed*: "Passionate grief does not link us with the dead but cuts us off from them." To turn a room into a shrine, to honor the death anniversary, to keep the dead one ever fresh and present in mind—paradoxically, this only distances us from the reality of the person who is lost to us. In a similar way, maybe, a highly passionate expression of grief in a memoir estranges the reader from the dead person and the mourner alike. Perhaps this is why I'm drawn most to the books that avoid a relentlessly raw, bewildered, stream-of-consciousness voice. And make no mistake, many grief memoirs can be described in those terms. Writing in the *Guardian* in 2011, Frances Stonor Saunders likens the grief memoirists to the hired mourners of the ancient Greek chorus, "renting their garments and generally disheveling." She slams the "metaphysical platitude, repetition, obsession, incoherence" in these books.

But it's not only Lewis who refuses the loud flail. Roger Rosenblatt's daughter collapsed and died on a treadmill at age thirty-eight, altering forever the lives of her husband, three children, two brothers, and parents. In *Making Toast*, Rosenblatt writes:

> Carl, John, and I had stood together on the deck in Bethesda the day after Amy died, and wept. Arms around one another, we formed a circle, like skydivers, our garments flapping in the wind. I could not recall seeing either of them cry since they were very young. I am not sure they had ever seen me cry, except on sentimental occasions. . . . The trouble with a close family is that it suffers closely, too. I stood with my two sons in the cold and put my arms around them, feeling the shoulders of men.

The phrase "shoulders of men" quietly conveys a world of hurt, and something more: We know that Rosenblatt now sees his sons as adults who, like he himself, must carry an adult grief.

In *Kayak Morning*, published two years later, Rosenblatt writes again of Amy and of grief. Asked why he wrote *Making Toast*, Rosenblatt explains in the later book that it was therapeutic, a way to keep his daughter alive. "When the book was finished," he writes, "it was as if she had died again." Would Lewis have cautioned Rosenblatt not to write that second book, because to let Amy go a little would bring her back with even greater force?

Oddly, then, grief memoirs may emerge from a consuming need to escape grief. The human mind may adapt to an overwhelming emotional experience by refusing to exile itself for too long in the darkest places. In *A Widow's Story*, Joyce Carol Oates writes:

> In my study, at my desk overlooking a stand of trees, a birdbath (not in use, in winter), a holly tree with red berries in which cardinals, chickadees and titmice bustle cheerily about, I am free to tell myself Ray would not be in this room with you anyway. Your experience at this moment is not a widow's experience.

But then the grief rebounds, and echoes, and echoes some more. It's inescapable, at least for a time, and the hardest part may be how cuttingly aware the mourner is of this fact. Lewis put it like this:

> Part of every misery is, so to speak, the mystery's shadow or reflection: the fact that you don't merely suffer but have to keep on thinking about the fact that you suffer. I not only live each endless day in grief, but live each day thinking about living each day in grief.

The character of Lewis's grief changes over time, and he is so brilliant at articulating this change that we derive insight and hope from his words. He's surprised to discover that one day he feels lighter, less closed off from God, and less agonized that the reality of Joy will leach away. The book's very slimness signals that even if grief doesn't end, its fiercest power fades.

Awareness of the weight of grief, and the changing topography of mental reflections about grief, are precisely what I believe other animals don't experience. And can animals feel guilt? In *Say Her Name*, Francisco Goldman's fictionalized account of the death of his wife, guilt seeps through the pages. Aura died in an accident while swimming with Goldman in waters off a Mexican beach. In describing his first meeting Aura, Goldman lingers over the young woman's beautiful face and eyes, her animated spirit. He recounts the greetings they exchanged: "Hola!" he says to Aura. "Hola," she responds. In this first movement toward her, he sets in motion an infinite chain of events that will come to encompass their love, their marriage, her death, and his grief. The chilling part comes when, in parentheses on the page, he imagines a conversation that never took place, coiled in the space between them as they met: "Hello! Meet your death," Goldman says. "Hello my death," Aura responds. In passages like these, grief memoirs convey the awful, grinding cost to our species of deep self-awareness.

Sometimes it's not guilt, or the knowledge of grief's enduring burden, but a sort of anticipatory grief that we feel. When the hands go cold at a doctor's grim expression, even as we wait for his words about a spouse, child, or friend, or when a loved one declines and we know only one outcome is possible, we take on board the coming loss, sometimes months or years before it happens. We anticipate the solitary path the dying person will navigate and envision our own lonely future. What will it be like, we wonder, on that day when we return alone to a home that will never be the same again? *The Rising*, the album Bruce Springsteen created after the 9/11 terror attacks in New York and Washington, DC, includes a song called "You're Missing":

> Pictures on the nightstand, TV's on in the den
> Your house is waiting, your house is waiting

But the listener knows that the house will wait forever. The song's title is the singer's refrain—*You're missing*—and the song ends with a terrible finality:

> God's drifting in heaven, devil's in the mailbox
> I got dust on my shoes, nothing but teardrops.

With 9/11, there was no time for anticipatory grief. Loved ones set out for work or to carry out the day's errands, and never returned.

Seen from this angle, it's clear that anticipatory grief may be a blessing as much as a burden; it allows us to put our love into words, and to prepare ourselves and others for the heart's coming absence. I felt both the blessing and the burden when, in the early 1990s, my friend Jim, only in his thirties, was dying of AIDS, at a time just before retroviral medications gave people with HIV an excellent chance to live with their disease. The funny thing about my relationship with Jim was, as more than one person remarked to us, that the English language lacks a kinship term for what we were to each other. "Friends" was accurate, but pale. We met in college, labored to find romantic love, and soon realized we were meant to share an intense platonic bond. Rooted in New Jersey, Jim followed me in my anthropologically mobile years, making visits to Oklahoma (graduate school), Kenya (field research), and Santa

Fe (dissertation writing). Then he became ill, and there was nothing to be done—and yet everything: exploring every medical option we could find, my traveling to him instead of him to me, a pledge near the end that I would think of him every day of my life. In the last days, I crossed that line from fervently hoping for a sick person's recovery to fervently wishing for a suffering person's death.

Other animals may alter their behavior when a companion is ill, as did the chimpanzees who surrounded a dying female at the Scottish safari park, or the goat who leaned hard against her friend the Shetland pony to help keep her on her feet. They may feel concern and act upon it. But only we look far ahead with dread, or relief, or a mix of the two, aware that death is coming. And when it arrives, and we mourn for another, we do so with a unique mix of private and public emotion, a balance that may even be adaptive for such a self-aware species. "When the living see that others lament the dead," writes Tyler Volk in *What Is Death?*, "they are consoled about their own future deaths."

Alone of all species, we may pour our lamentations into art, as grief-memoir writers do. With the exception of the embodied grief that may be expressed in dance, though, it may be when we still our unique creativity that we feel closest to other animals who grieve. We grieve with human words but animal bodies and animal gestures and animal movements.

15 THE PREHISTORY OF GRIEF

When they died, the boy was no older than twelve or thirteen, the girl no older than ten. The boy apparently had developed normally, but the girl showed signs of a bilateral deformity of the femurs, meaning that her legs were short and curved; she walked with a bowed gait. The children lived in a settlement we now call Sunghir, along a riverbank in Russia about two hundred kilometers east of today's Moscow. Sunghir's permafrost attests to a challenging climate. When it came time to dig through the cold earth to lay the children's bodies to rest, the Sunghir community drew together. Through a collective eye for beauty and many hours of skilled labor, these people ensured that the children would leave this world by way of a spectacular burial ritual.

We have no eyewitness reports to the ceremony, because the children died twenty-four thousand years ago. This period of the Paleolithic predates not only writing but also settled village life and the domestication of crops or most animals. This isn't to say that the Sunghir people, anatomically modern *Homo sapiens*, led simple lives. Gorgeously rendered animal images, alive with color and painted on walls at caves like Chauvet in France starting around thirty-five thousand years ago, reveal the cultural complexity of our *Homo sapiens* ancestors.

Archaeologists' descriptions invite us to imagine that long-ago day when the Sunghir community convened at the grave. Vincenzo Formicola and Alexandra Buzhilova write:

> The two children were buried head to head in supine position in a long, narrow, and shallow grave dug into the permafrost. The skeletons were covered with red ocher and accompanied by extraordinarily rich and unique grave goods. Thousands of ivory beads, probably sewn onto clothes, long spears of straightened mammoth tusks (one of which is 240 cm long), ivory daggers, hundreds of perforated arctic fox canines, pierced antler rods, bracelets, ivory animal carvings, ivory pins, and disc-shaped pendants were part of the ornamentation of the burial.

In the world of anthropology, this description of Sunghir's double-child burial is famous. It was a rare practice to bury children so long ago, at least judging from the graves that archaeologists have to this point uncovered. Even rarer was the girl's deformity, but it strengthens scientists' suspicion that prehistoric burial for this age range occurred more often when the child's anatomy deviated from the normal. Still, only one of the Sunghir children fits into this category, and all indicators are that she died for reasons unrelated to the bowing of her legs. Archaeologists feel certain the two deaths happened close enough in time for the children's burials to be simultaneous. Perhaps an accident befell the boy and the girl when they were foraging or carrying out some other activity on behalf of the community, or maybe they fell victim to disease.

The compelling nature of the bones and artifacts from this site explains part of Sunghir's notoriety, but I think there's more to it. Can we fail to feel a connection to these people, so distant in time, when we learn of the actions they took in the face of death? The detail that catches in my throat comes from the archaeologists' report: thousands of ivory beads, probably sewn onto the children's clothes. Faced with unrelenting challenges to their very survival in this cold climate, these hunter-gatherers took the time to decorate the young bodies before burial. For me, the sewn beads are the Sunghir people's grief made material.

Of course, it's possible that I am wrong and that the hard work of the Sunghir burial preparations went on in the absence of mourning. But here is where the stories in this book may aid in attempts to reconstruct our past. A variety of highly social birds and mammals show the capacity for grief—including corvids, geese, dolphins, whales, elephants, gorillas, and chimpanzees. If I'm right, and individual animals from these species mourn because they've felt love for another creature, can it be such a stretch to suggest that love and grief were expressed by some individuals of our own species twenty-four thousand years ago? Wouldn't these emotions be a likely by-product of close community living in a smart, social, and self-aware primate?

While the expression of grief spans both time and species, the practice of community burial *isn't* known from nonhuman animals and is rare even in our own lineage. From the time our ancestors first stood upright over four million years ago, through their first crafting stone tools around two and a half million years ago and the onset of big-game hunting somewhere between two million and one-half million years ago, no sign survives of the burial or cremation of the dead. This fact carries fascinating implications when we consider the large numbers of individual involved. A demographic research group estimates that 107 billion people have lived and died between about fifty thousand years ago and the present. I'm suspicious of endorsing any precise figure, because calculations of this nature involve messy assumptions and rough guesses about population numbers in our past. As a thought experiment, though, this exercise makes a point. Correcting for the fact that our lineage began not fifty thousand but around six million years ago, we see that a vast number of humans (and human ancestors) have been born, lived, and died. What happened to their bodies? Did anyone mourn the dead? When did a social, ceremonial response of mourning for the death of the individual originate?

Sunghir gives us a fixed point in time by which hunter-gatherers (at least some hunter-gatherers) carried out burial ceremonies, probably with attendant emotion. Using Sunghir as a starting point and working backward, is it possible to uncover archaeologically the origins of grief in the human lineage?

In Israel, two prehistoric cave sites offer treasure troves of information about how *Homo sapiens* lived around a hundred thousand years ago. At Qafzeh in the lower Galilee region, and at Skhul at Mount Carmel, early modern people carried out the earliest known intentional burials. (The Qafzeh burials date to about 92,000 years ago; the Skhul dates could be anywhere from 80,000 to 120,000 years ago). Nowhere near as elaborate as those at Sunghir, the burials at Qafzeh and Skhul show unmistakable signs of deliberate care for the dead amid a thriving culture.

Archaeologist Daniella E. Bar-Yosef Mayer and her colleagues describe the Qafzeh culture as consisting of people who decorated their (living) bodies with red ocher, collected shells during trips to the seashore (about forty-five kilometers away), and applied red ocher to some of the shells in what may be an early example of a sort of artistic manipulation. Both children and adults were buried at the cave site; in one case, an adolescent was interred with an antler on the chest. At Skhul, an individual was buried with a boar's jaw; shells were intentionally perforated and included in some of the graves.

These Israeli sites offer an evolutionary foundation for humans' careful treatment of their dead, and hint at what we know came afterward. Sometimes, scholars of the origins of religion force a link between the presence of special goods in the grave and a cultural belief in the afterlife, but there's no reliable way to correlate the two. Grave goods could just as easily be an indication of respect and love for the dead as of community-held beliefs about what happens after death. (You'll notice I have made no argument for belief in the afterlife, or the presence of religious ritual, at Sunghir.) I think Tyler Volk is right, though, to link human death rituals with individuals' reflections on their own mortality. When people come together around a body, Volk writes in *What Is Death?*, it "forces them to face death. . . . Death serves to awaken the consciousness of the living."

As *Homo sapiens* flourished and some people began to farm, humans' patterns of meaning-making around death changed. A man-and-lamb double burial beneath a house floor at Çatalhöyük, Turkey, around eight thousand years ago suggests an emotional relationship between humans and the animals they domesticated. The grand-scale tombs of ancient Egypt were filled, a few thousand years later, with food intended for people to eat in the afterlife. A chronology of prehistoric practices shows that the human imagination became increasingly attuned to matters of death, and life after death.

Even early on, though, *Homo sapiens'* death practices were concerned with the symbolic, not just the functional. At Qafzeh and Skhul, red ocher became a tool of cultural expression, as it did for prehistoric peoples elsewhere. Rich in iron, deep red in color, ocher played a major role at Blombos Cave in South Africa, a "go-to" site for understanding the lives of early *Homo sapiens*. Coastal residents, the Blombos people made good use of marine resources. They speared fish, hunted seals and dolphins, and gathered periwinkles. The old idea that a "revolution" in modern human behavior occurred only thirty-five thousand years ago in Europe can still be found in some textbooks, but remarkable discoveries at Blombos have firmly falsified this view.

The Blombos people created paint pigments by using hammerstones and grinding stones in smart ways. We know this because archaeologist Christopher Henshilwood and his team discovered an artist's studio at Blombos dated to a hundred thousand years ago (the same time period as Qafzeh and Skhul farther north). The Blombos hunter-gatherers ground hard ocher into powder, sometimes mixing it with charcoal and with oil from seal bones. Abalone shells became dual-purpose tools, serving both as mixing bowls and as containers for the resulting pigments. Henshilwood's detective work takes us right to the verge of an exciting view of the ancient artists' lives, but the artifacts remain silent as to how our ancestors used the pigments they created. Did they color their tools? Paint images on walls? Did they apply pigments to their own bodies, as the Qafzeh and Skhul people did around the same time?

Blombos was home to early *Homo sapiens* for many thousands of years. Around seventy-five thousand years ago, its residents incised blocks of red ocher with patterned marks. While this isn't writing, such patterns can only be generated by a mind that thinks abstractly and breaks free of an exclusive focus on day-to-day survival skills. Jewelry-making requires this ability as well; the Blombos people pierced the shells of small mollusks in precise ways that, together with wear patterns on the shells, shows they were worn in self-ornamentation. It's tempting to conclude that the dead must have been buried at Blombos as they were at the two Israel sites, but no burials have so far been uncovered there.

The emerging picture of life for early *Homo sapiens* in Africa and the Middle East is one of creative self-expression by people who thought about their lives—and felt their lives. A picture that is similar in certain ways is emerging for our close cousins the Neandertals. At around thirty thousand years ago, populations of the big-brained, robust-bodied Neandertals went extinct, except in the sense that elements of the Neandertals' genetic lineage do live on in some modern populations. For thousands of years before that point, Neandertals had coexisted with anatomically modern humans, and at certain times and places (though never in Africa, where they did not reside), they directly met up with our species.

It would be a serious mistake to view these people through the old "caveman" stereotype that portrayed them as shambling, club-carrying creatures. With spear-wielding skills, Neandertals hunted dangerous big game like mammoth. Some Neandertals modified teeth of bear, wolf, and deer to wear as pendants, or adorned the molar of a mammoth with red ocher, smoothing and polishing the tooth as a kind of symbolic keepsake. And some buried their dead. At the site of La Ferrassie in France, Neandertals covered the body of a group member with a limestone slab; at Teshik-Tash in Uzbekistan, they encircled the body of a child with goat horns.

Careful burials of the dead are not known from prior evolutionary periods, but one site may be telegraphing clues to archaeologists about treatment of the dead by earlier ancestors. At Spain's "Pit of the Bones" (La Sima de los Huesos), the remains of thirty-two individuals cluster together at the bottom of a forty-five-foot shaft. The date? Three hundred thousand years ago. Could the Sima people have deposited the bodies in the shaft as an act of respect or veneration? Or were the bodies sent tumbling down by an act of aggression or malice? Perhaps there was no deliberate act at all, and the bodies somehow fell into the shaft accidentally. The site yields no answers. Further back in time than this, the material evidence gives no clues at all about the death practices of our ancestors.

For many people, "Lucy," who lived three million years ago in the Rift Valley, is a touchstone for understanding the human family tree. Famously discovered by Don Johanson in Ethiopia almost forty years ago, Lucy and her kind (*Australopithecus afarensis*) strode upright through a woodland ecosystem teeming with other mammals and birds. Lucy died at about the age of twenty. Gradually, her body skeletonized—reduced to bone—in the spot where she died, just as happens with wild animals today (unless their carcasses are consumed or transported away by other animals).

Fascinating as these glimpses of the origins of human mortuary practices may be, no site excavation, grave-good inventory, or bone analysis can reveal what emotion was felt by the family or community of someone who died thousands of years ago. Yet, as I have already argued, the weight of animal-grief stories collected in this book supports the likelihood of grief in our own prehistory. What the artifacts and bones don't tell us, and what scientists of the past may be reluctant to speculate about, comes clearer for us in comparative context. Twenty-four thousand years ago at Sunghir, even a hundred thousand years ago at Qafzeh, Skhul, and Blombos, our ancestors had the cognitive and emotional resources to feel grief and the community structure to support its expression. It is only emotional *capacity* that this comparative context illuminates, though, and we would do well to keep in mind that the capacity to experience an emotion doesn't always result in the *expression* of that emotion.

For some of us today, death is finality: life ends at the moment we cease breathing. For others of us who have a transcendent belief in the soul and its timeless continuity, the death of a fleshly body doesn't equal the death of the person. Believers in a sacred afterlife or in reincarnation may regard death as the passage to an existence that is far more fulfilling. When death is not viewed as the end of meaningful existence, mourning may be tinged with a hint of celebration.

Human meaning-making around the body, death, and mourning in the modern world is infinitely complex, and meaning-making about the body, death, and mourning in the past remains elusive. Anthropology can bring us no closer to a prehistory of grief than to document the elaborate care with which some bodies were buried, and to strongly suggest, based on examples from nonhuman animals, that such acts of care reverberated with feelings of loss.

In chapter 14, I hammered away at the "uniquely human" perspective on grieving, making the point that only our species turns mourning into art. In this chapter, I've outlined prehistoric human rituals of burial that are unequalled in their elaborate nature by the actions of any other animal. At the same time, I'm appealing here to the emotional capacities of other animals to argue that, in at least some places at some periods of the past, our extinct ancestors carried out those elaborate rituals in a state of felt grief. In this way, I've returned to that balancing act I mentioned in the prologue—the need I feel as an anthropologist to acknowledge how our species differs from others in grief behavior even while I put most of my effort into highlighting the points of cognitive and emotional similarity with other creatures.

On a recent visit to Berlin, I felt the full force of the uniquely human response to death. To walk among the 2,711 concrete slabs at the Memorial to the Murdered Jews of Europe is a disordering experience. One block from the Brandenburg Gate, on an open-air site that may be visited around the clock, the stelae line up in parallel rows of varied height. Moving up and down the rows, glimpsing another person here and there at some intersection on the path, I felt exactly as I imagine the architect intended I should: surrounded by indifferent sameness, thrown back on myself, overwhelmed by silence and that sense of disorientation. *How* those concrete slabs made me feel that way I am at a loss to articulate, but they did. Beneath the slabs, in an underground exhibit, are the names of every Jew known to have been murdered during the Holocaust, along with many photographs and text passages. The images and words are haunting, but in that space—arranged as in a conventional museum—my experience was ordered, familiar, and thus much different in nature from what I felt wandering among the stelae.

At the Berlin memorial, it's concrete slabs. In Oklahoma City, it's 168 chairs laid out in neat rows. In Lower Manhattan, it's two open spaces, surrounded by trees and cascading water, that mark the empty footprints of the World Trade Centers' Twin Towers. In Hiroshima, it's the statues, bridges, open areas, and beautiful clock tower of the Peace Memorial Park. In Kigali, Rwanda, it's 250,000 bodies interred on the grounds of the Genocide Memorial Center. In the aftermath of a blinding flash, a catastrophic day, or war's grinding attrition, our mourning becomes global in a way never before possible in our history or prehistory.

At this scale, grief spreads across space and time like waves across the sea. After the writer Francisco Goldman's young wife, Aura, was killed by a wave while swimming at a Mexican beach, Goldman felt compelled to explore the behavior of waves. Waves, he later wrote,

> travel across the ocean in sets, or trains, and it's never just one train that arrives at a beach, because along the way wave-trains meet or converge or overtake one another and mix, older waves with somewhat younger

ones. But even a moderate wave, I've since learned, breaks and surges toward the shore with the innate force of a small automobile going at full throttle.

And so it is with the breaking and surging of response to mass-scale death. It ripples out from immediate survivors to extended families, from the local community to the whole nation, across continents and oceans. The grief of one converges with the grief of many, tumbling into and sometimes exacerbating the felt emotion. This upwelling is haunting, and entirely human.

A generation of Americans witnessed this process in the days, months, and years after the 9/11 terrorist attacks. It's a cliché that many people of many nations remember with startling clarity precisely where they were and what they were doing on that Tuesday morning. I had begun teaching 125 anthropology students at 9:30 a.m., and as the news from Manhattan and the Pentagon became increasingly dire and the students and I became increasingly anxious, I ended class early. Our mourning began that very day, but where had it been the day before, on Monday, September 10? Was it gathering itself in baby ripples, conjoining to explode the next day with terrible force? The question may sound peculiar, but in the context of Goldman's study of waves it makes sense to me. Goldman muses about the long journey of "Aura's wave," the strong pull of water that eventually tumbled her so violently through the surf that she would die. Most surface waves travel thousands of miles before breaking on a beach. "It's not the water itself that travels, of course, but the wind's energy," he writes. "Large waves charge steadily along on high-velocity winds that have been traveling across the open ocean for many thousands of miles and for days."

It wasn't, I think, grief that gathered itself together in the days and hours before 9/11. It was love: the love that people felt as they kissed or waved good-bye to family and friends that morning. The love drives the grief as the wind drives the ocean waves.

When the first tower collapsed in Manhattan that day, Jean-Marie Haessle, a French-born artist, began to hurry uptown. But he stopped on Wall Street and scooped up some of the dust that was falling all around him. It was an impulse, he told the *New York Times*. The dust calls to mind, he said, his own eventual death; he keeps it safe in the paper envelope in which he first collected it. What makes up this dust? Surely it must contain parts of the fallen tower, bits of the office machines and papers and other compressed objects from everyday working lives. But does it contain more . . .? I find it impossibly painful to ask the question any more precisely; we all remember the thousands lost, and realize what other material the dust may contain. Haessle curates the dust for a museum audience of one, and for him it carries great symbolic power.

I see an invisible ribbon of time that connects Haessle, the contemporary artist in New York, with our forebears from Sunghir in Russia and Qafzeh and Skhul in Israel. To set aside space for the dead, to mark the relationship of the living with the dead through an elaborate burial or a respectful keeping of ashes—or through a capital city's disorienting memorial that draws millions of visitors from around the world—is at one and the same time a thoroughly human act and an act that is possible because we are social animals who evolved from other social animals who grieve.

AFTERWORD

"[Grief] occurs widely in other social mammals and in birds, for example after loss of a parent, offspring or mate." So wrote John Archer on the first page of *The Nature of Grief.*

It's rare to find such a complete embrace of animal grief in the social-science literature—especially in 1999, when Archer was writing, before the current wave of scientific interest in mammal and bird mourning. Archer followed up his straightforward assertion with only a three-page review of evidence that supports it. He discusses corpse-carrying in monkeys and apes, anecdotal reports of grief in birds and dogs, and results of "separation experiments" showing that the young in a variety of species become distressed when separated from their mothers. Of course, Archer could not have included information on animal mourning from the last fifteen years. It's safe to say that science-minded readers may find a gap between Archer's confident claim of animal grief and the support he marshals for it from the animal world.

Have the stories presented in these pages succeeded in closing the gap between a claim for animal grief and the evidence? Unsurprisingly, my own verdict is "yes," but I know that it is important to make distinctions between the strong and the moderate or weak evidence contained in these pages. In deciding between these alternatives, one benchmark could be the ideal definition of grief that I offered in the prologue: Grief can be said to occur when a survivor animal acts in ways that are visibly distressed or altered from the usual routine, in the aftermath of the death of a companion animal who had mattered emotionally to him or her.

Using this benchmark, numerous examples in this book offer strong evidence for grief among animals living in the wild. Long-term research on two elephant populations in Kenya—Samburu in the north and Amboseli in the south—has tracked individual elephants' response to death. Kin and friends reacted to the death of the elephant matriarch Eleanor at Samburu in distressed or unusual ways, and elephants in Amboseli caressed the bones of their matriarch. Interestingly, one piece of evidence from Samburu complicates the framework I'm using. Females mourned for Eleanor who *hadn't* been particularly close to her in life, leading researcher Iain Douglas-Hamilton to posit a "generalized" response to death among elephants. If Douglas-Hamilton is right, elephants may show more community-wide (in addition to kin- and friend-based) emotional responses to death than other animals—or perhaps in the coming years we may find community-wide reactions in other species.

The elephant evidence provides, I believe, the strongest case for animal grief in the wild, closely matched by that for dolphins, chimpanzees, and some birds. In dolphins, the maternal response to infant death is heart-wrenching to watch, attesting to severe maternal distress. Intriguingly, corpse-carrying mother chimpanzees (and monkeys) don't to my knowledge show overt emotion, but some chimpanzees do demonstrably mourn, as we know from Flo's son Flint at Gombe and Tina's little brother Tarzan at Tai, Cote d'Ivoire. With pair-bonded birds, mourning may lead to serious depression in the survivor.

Not all of my examples from the wild convincingly meet the strict definitional criteria. With the male sea turtle in Hawaii who had lost Honey Girl, his presumed mate, the bison in Yellowstone National Park who inspected their companion's carcass, and the corpse-carrying monkey mothers who seem unaffected emotionally by their burden, the evidence is suggestive of grief to varying degrees, but not conclusive. Even in these unresolved cases, however, the absence of a family member, a groupmate, or a social partner changed the survivors' behavior in measurable ways. And in the case of monkeys, we have the baboon anecdote from Okavango in which the mother Sylvia grieved for her daughter Sierra, plus the physiological data showing a chemical signature of bereavement in multiple individuals.

Among animals who live closely with humans, in homes, farms, sanctuaries, or zoos, some cases of grief also meet my stringent definition. I can't imagine interpreting the stories of cat sisters Willa and Carson, or dog friends Sydney and Angel, with a focus other than grief: to my mind, the signs of love and mourning in these stories are too overwhelming to reasonably be interpreted away. The same is true for rabbits and horses, in numerous examples that I have shared.

The two rescued mulard ducks Kohl and Harper made an especially poignant pair; that Harper loved and grieved for his friend is, for me, beyond sensible contesting. The behavior of Tarra the sanctuary elephant when her small dog friend Bella died reminds us that two animals of very different natures may experience keen friendship and, for the survivor, sadness when that friendship comes to a sudden end.

Zoos may become a leading source of data on animal grief in the future. Right now, the recorded behaviors of gorillas and chimpanzees around death in zoos and similar captive institutions raise more questions than they answer. When the female chimpanzee Pansy died in a Scottish safari park, why did the male Chippy attack her corpse? What does it signify when the companions of a zoo gorilla who has died continue to search for that individual even though the body had been visible to them? Perhaps more than any other animals, our closest living relatives, the African apes, clue us in to the great variability in mourning behaviors, both in wild populations (in the case of chimpanzees) and captive ones.

And when it comes to unanswered questions, the whole issue of animal suicide stands out. The examples of bear and dolphin emotional suffering I've included take us into the arena of possible intense grief for a lost loved one or for one's own intolerable living situation. Science has barely even considered these possibilities, or, if animal suicide does exist, its range of possible causes.

We humans have our own species-specific ways of mourning. The final two chapters have considered how we may turn grief into art, and how our burial rituals and other death practices have evolved over many millennia. Yet what stands out for me isn't human uniqueness, but the knowledge that animals other than humans do love, and do grieve. As

I have stressed throughout the book, this statement shouldn't become a litmus test of emotional complexity for other species. Some dogs will grieve, depending on their personalities and the contexts in which they live. Some dogs won't. The same is true for chimpanzees and other species who have mourned in a way that we recognize. The expression of animal emotion doesn't lend itself well to bold generalization across individuals—any more than the expression of human emotion would.

In reviewing the stories of animal grief in order to write these closing passages, it comes back to me how thoroughly joy tangled with sadness as I researched and wrote. The sadness emerged, of course, because I immersed myself in the lives of animals who had a deep channel of grief running through their emotions, sometimes briefly and sometimes for extended periods.

But then there was the joy: I discovered the depth of animal love. Because of this, I look at many animals differently now than I did even three years ago. Compared to my other books, this one required a broader look across the animal kingdom; the payoff came when I discovered emotions in animals—in farm animals most of all, but really across the board—that were more complex than I had ever suspected.

Also joyful was my experience of friends, relatives, scientific colleagues, and complete strangers who know me only through my writing, all sharing stories about how animals grieve (or in some cases, don't grieve). A sense of "being in it together," of wanting to figure out new ways of perceiving, describing, and analyzing animal love and animal grief, knotted us together.

Views contrary to my own were at times enlightening. When I wrote about animal love at NPR's 13.7 blog, I didn't question whether animals feel love but asked, how do we recognize the love that other animals feel? Even as ideas and examples poured in, some readers insisted we shouldn't pick apart animal love. Trena Gravem asked, "Why define love? Why overthink it and try to analyze it? Instead we should be extremely thankful that we have love for others, and they for us; and of course it goes without saying, this includes animals. Sadly, only children do not question this." Meg Ahere wrote, "I would prefer to start from the assumption that every animal feels emotions and loves others in its own way."

These views are articulate and embrace a stance that is wide-open to the expression of complex animal emotion. Yet for me, as a scientist, the bottom line is this: If every animal who acts in a positive or compassionate way toward a companion is said to love, and if every animal who responds with some display of emotion to a dead companion is said to grieve, we run the risk of diluting the phenomenon we want to understand. And we don't learn much.

I hope that the ideas and questions in this book will be taken up by others who will strive to discover more about how individual animals grieve, or don't grieve. Maybe my very definitions of love and grief can be improved, and entirely new questions added to the mix. What's important is to continue the conversation, because it's not one mired only in theoretical concern, or even in concerns about how we may understand ourselves better by understanding other creatures. To plumb the depths of animal thinking and feeling means to reassess how we, collectively as a society and individually as persons, treat other animals. I've already discussed the beneficial practice of allowing surviving animals some time with the bodies of their loved ones. With this practice, we recognize that animals think and feel, and offer to grieving animals the compassion and dignity they deserve.

And here come echoes of my joy-and-sadness theme again: For animal lovers, the knowledge weighs heavy that animals in the wild, housed on farms or in sanctuaries or zoos, or alongside us in our homes, may struggle or may have struggled in various ways because of human neglect or abuse. Even here, though, there's room for joy: We may bring about a shift, a sea change from treating animals as *somethings* to treating them as *someones*—just as Farm Sanctuary teaches us.

I would like to conclude on a personal note. In 2005, a column called "Always Go to the Funeral" was broadcast by NPR as part of its "This I Believe" essay project. In it, Deirdre Sullivan described her parents' insistence that she, as a shy teenager, attend the funeral of a grade-school teacher. Trying to squeeze out a few words of condolence to the teacher's family, the young Sullivan felt mortified. Only later did she appreciate being raised to understand that some acts mean so much to others that your own discomfort, or inconvenience, matters little. She concludes with this passage:

> On a cold April night three years ago, my father died a quiet death from cancer. His funeral was on a Wednesday, middle of the workweek. I had been numb for days when, for some reason, during the funeral, I turned and looked back at the folks in the church. The memory of it still takes my breath away. The most human, powerful and humbling thing I've ever seen was a church at 3:00 on a Wednesday full of inconvenienced people who believe in going to the funeral.

Also on an April night, my own father died. It was 1985, and he had lived to be sixty. First in the Navy during World War II, later as a fireman and then for decades as a New Jersey state policeman who fought organized crime, he served others. During his funeral, the gunfire salutes offered by his former state police colleagues brought tears to my eyes. What is rooted most firmly in my heart since that day isn't the official ceremony, though. It is the gathering together of many people who interrupted their spring day to sit with us, to honor my father with their words, and to transfer their strength to my mother and me.

It's no accident, I think, that I chose to write a book about grief when I entered my mid-fifties. It's true that in doing research for earlier works, I kept bumping up against bits and pieces of evidence for animals' emotional responses to death. In that sense, this book grew naturally from seeds planted in the previous two. Yet there's more going on. I am part of the great wave of baby boomers who now approach—or have reached—the retirement years. My only child is in college. My mother is in assisted living. With a narrow escape after emergency surgery at the age of eighty-four, she finds herself needing more complicated care than ever before. Now eighty-six, she may live as long as her own mother, to one hundred, or she may be gone sooner. My mother's life is intertwined with mine in a way it hasn't been before—except, of course, when I was very young. When I talk with friends around my age, our conversation veers often into elderly-parent territory. We share the worry, exhaustion, and yes, the satisfaction too, of caring in various ways for our mothers and fathers.

As I negotiate details of my mother's stays in hospitals, nursing-home rehabilitation centers, and her assisted-living residence, I feel profound love for her mixed with an anticipatory grief. Frequently, I learn that someone close to me is in the grip of fully realized grief: One friend's mother dies shortly before her ninetieth birthday, after a long struggle with cancer. Another's father, in his eighties, is gone after a short period of intense physical decline; my friend is sure he willed himself to die, helped along by his refusal to eat. Another friend's son dies right after Christmas in a terrible car wreck at age seventeen. For that mother, I feel wild sorrow, and know of no way to offer comfort. All I do know is to share her love for her son, which survives him in abundance.

It won't ease our deepest grief to know that animals love and grieve too. But when our mourning becomes a little less raw, or is so far only anticipated, may it bring genuine comfort to know how much we share with other animals? I find hope and solace in the stories in these pages. May you find hope and solace in them as well.

ACKNOWLEDGMENTS

My first and heartfelt thanks go to the people who have met or communicated directly with me for this book about the animals they live with, or once lived with: Karen and Ron Flowe, Nuala Galbari, Janelle Helling, Charles Hogg, Connie Hoskinson, David Justis, Melissa Kohout, Jeane Kraines, Michelle Neely, Mary Stapleton, and Lynda and Rich Ulrich.

I am grateful also to those who wrote comments about their experiences with animal grief in response to my posts at NPR.org's *13.7 Cosmos & Culture* blog. To my 13.7 editor, Wright Bryan, thanks for teaching me so much.

To the scientists and zoo staff members who generously responded to my questions and shared material with me, I owe sincere thanks: Karen Bales, Tyler Barry, Marc Bekoff, Melanie Bond, Ryan Burke, Dorothy Cheney, Jane Desmond, Anne Engh, Sian Evans, Peter Fashing, Diane Fernandes, Roseann Giambro, Liran Samuni, Karen Wager-Smith, and Larry Young.

My admiration as well as my gratitude goes to the staff of the Elephant Sanctuary–Tennessee, the Farm Sanctuary, and the House Rabbit Society, who helped me with material about animal mourning and who help thousands of animals in need.

Through its research leave program, the College of William and Mary made possible the period of intense reading and writing from which this book emerged. To provost Michael Halleran, director of research communications Joseph McClain, and my colleague anthropologist Danielle Moretti-Langoltz, special primate gestures full of thanks.

I have fond memories of sitting around a table in the Levine-Greenberg literary agency in Manhattan years ago, brainstorming about this book (as yet only an idea) with Jim Levine and Lindsay Edgecombe. Lindsay and Jim believed in the ideas behind the words "how animals grieve" and helped tremendously in shaping the book as it came to fruition.

As the book went to press, Jill Kneerim of the Kneerim-Williams Agency provided superb guidance and support, and turned me firmly toward my future of writing about animal emotion.

All along, the experience of working with people at the University of Chicago Press has been an intellectual treat and a personal delight. Christie Henry always says the *best* thing at the *best* time and, through her editorial insights, has made this a better book in at least ten different ways. Also at the Press, Levi Stahl, Joel Score, and Amy Krynak have been so good to me, and to the book too, that I send them a sincere thank you.

To Stuart Shanker, how adequately to acknowledge many years of shared work projects, mutual support, and exchanged tales of children, chickens, and cats? Please know your friendship matters so much, every day.

I've succumbed to convention in mentioning those closest to my heart last. In one sense, it's a small family, consisting of my husband Charles Hogg, my daughter Sarah Hogg, and my mother Elizabeth King. To my mother, thank you for everything you have given me, including a ton of books from the earliest years onward, and a love of reading. To my daughter, I will always cherish our serious and silly talks about animals (including Sir Lancelot!), about writing, and about standing up for what (and who) we care about. To my husband, I can only say, you're amazing to me. I've loved you since that first fateful fall of 1989, but I love you more by the day as I see with new eyes the depth of your commitment to animals.

And yet, those are just the *Homo sapiens*! Our family circle is larger. To all the animals who have loved me back over the years (cats, dogs, rabbits) and to all the others who have regarded me with a mild friendliness or outright indifference but who are still gorgeous and still loved (monkeys, apes, bison, frogs, birds, and more), trying to enter your emotional worlds is a joy and a responsibility and I hope I've gotten some important parts right.

READINGS AND VISUAL RESOURCES

PROLOGUE

Bekoff, Marc. "Animal Love: Hot-Blooded Elephants, Guppy Love, and Love Dogs." *Psychology Today* blog, November 2009. http://www.psychologytoday.com/blog/animal-emotions/200911/animal-love-hot-blooded-elephants-guppy-love-and-love-dogs.

Kessler, Brad. *Goat Song: A Seasonal Life, a Short History of Herding, and the Art of Making Cheese*. New York: Scribner, 2009. Quoted material, p. 154.

Krulwich, Robert. "'Hey I'm Dead!' The Story of the Very Lively Ant." National Public Radio, April 1, 2009. http://www.npr.org/templates/story/story.php?storyId=102601823.

Potts, Annie. *Chicken*. London: Reaktion Books, 2012.

Rosenblatt, Roger. *Kayak Morning*. New York: Ecco, 2012. Quoted material, p. 49.

CHAPTER ONE

Coren, Stanley. "How Dogs Respond to Death." With a sidebar by Colleen Safford. *Modern Dog*, Winter 2010/2011, 60–65. Quoted material, p. 62.

Harlow, Harry F., and Stephen J. Suomi. "Social Recovery by Isolation-Reared Monkeys." *Proceedings of the National Academy of Sciences* 68 (1971): 1534–38. Quoted material, p. 1534. http://www.pnas.org/content/68/7/1534.full.pdf.

King, Barbara J. "Do Animals Grieve?" http://www.npr.org/blogs/13.7/2011/10/20/141452847/do-animals-grieve.

Renard, Jules. *Nature Stories*. Translated by Douglas Parmee. Illustrated by Pierre Bonnard. New York: New York Review of Books, 2011. Quoted material, p. 39.

CHAPTER TWO

Coren, Stanley. "How Dogs Respond to Death." With a sidebar by Colleen Safford. *Modern Dog*, Winter 2010/2011, pp. 60–65.

Dosa, David. *Making Rounds with Oscar: The Extraordinary Gift of an Ordinary Cat*. New York: Hyperion, 2010.

Hare, Brian, and Michael Tomasello. "Human-Like Social Skills in Dogs?" *Trends in Cognitive Science*, 2005. http://email.eva.mpg.de/~tomas/pdf/Hare_Tomasello05.pdf.

King, Barbara J. *Being with Animals*. New York: Doubleday, 2010.

Zimmer, Carl. "Friends with Benefits." *Time*, February 20, 2012, 34–39. Quoted material, p. 39. (For responses by Patricia McConnell, see http://www.patricia mcconnell.com/theotherendoftheleash/tag/carl-zimmer.)

VIDEO Ceremony to honor the dog Hachiko, Tokyo, April 8, 2009. One can see the statue of Hachi in the opening frames. (In Japanese.) http://www.youtube.com /watch?v=ffB6IEFsD9A.

VIDEO Heroic dog rescue on the highway in Chile. http://today.msnbc.msn.com /id/28148352/ns/today-today_pets_and_animals/t/little-hope-chiles-highway -hero-dog/.

PHOTO Hawkeye the dog at Jon Tumilson's casket. http://today.msnbc.msn.com /id/44271018/ns/today-today_pets_and_animals/t/dog-mourns-casket-fallen -navy-seal/.

CHAPTER THREE

Farm Sanctuary, "Someone, Not Something: Farm Animal Behavior, Emotion, and Intelligence." http://farmsanctuary.wpengine.com/learn/someone-not -something/.

Hatkoff, Amy. *The Inner World of Farm Animals*. New York: Stewart, Tabori & Chang, 2009. Quoted material, p. 84.

Marcella, Kenneth L. "Do Horses Grieve?" *Thoroughbred Times*, October 2, 2006. http://www.thoroughbredtimes.com/horse-health/2006/october/02/do-horses -grieve.aspx.

CHAPTER FOUR

Archer, John. *The Nature of Grief: The Evolution and Psychology of Reactions to Loss*. New York: Routledge, 1999.

House Rabbit Society. "Pet Loss Support for Your Rabbit." http://www.rabbit.org /journal/2-1/loss-support.html.

Wager-Smith, Karen, and Athina Markou. "Depression: A Repair Response to Stress-Induced Neuronal Microdamage That Can Grade into a Chronic Neuroinflammatory Condition." *Neuroscience and Biobehavioral Reviews* 35 (2011): 742–64.

CHAPTER FIVE

Bibi, Faysal, Brian Kraatz, Nathan Craig, Mark Beech, Mathieu Schuster, and Andrew Hill. "Early Evidence for Complex Social Structure in *Proboscidea* from a Late Miocene Trackway Site in the United Arab Emirates." *Biology Letters* (2012). doi: 10.1098/rsbl.2011.1185.

Douglas-Hamilton, Iain, Shivani Bhalla, George Wittemyer, and Fritz Vollrath. "Behavioural Reactions of Elephants towards a Dying and Deceased Matriarch." *Applied Animal Behaviour Science* 100 (2006):87–102.

Elephant Sanctuary. "Tina." http://www.elephants.com/tina/Tina_inMemory.php.

Gill, Victoria. "Ancient Tracks Are Elephant Herd." BBC, February 25, 2012. http://www.bbc.co.uk/nature/17102135.

McComb, Karen, Lucy Baker, and Cynthia Moss. "African Elephants Show High Levels of Interest in the Skulls and Ivory of Their Own Species." *Biology Letters* 2 (2005): 2–26.

Moss, Cynthia. *Elephant Memories: Thirteen Years in the Life of an Elephant Family*. New York: William Morrow, 1988. Quoted material, p. 270.

VIDEO Amboseli elephants' response to a matriarch's bones: http://www.andrews-elephants.com/elephant-emotions-grieving.html.

CHAPTER SIX

Bosch, Oliver J., Hemanth P. Nair, Todd H. Ahern, Inga D. Neumann, and Larry J. Young. "The CRF System Mediates Increased Passive Stress-Coping Behavior Following the Loss of a Bonded Partner in a Monogamous Rodent." *Neuropsychopharmacology* 34(2009): 1406–15.

Cheney, Dorothy L., and Robert M. Seyfarth. *Baboon Metaphysics: The Evoution of a Social Mind*. Chicago: University of Chicago Press, 2007. Quoted material, pp. 193, 195.

Engh, Anne L., Jacinta C. Beehner, Thore J. Bergman, Patricia L Whitten, Rebekah R Hoffmeier, Robert M. Seyfarth, and Dorothy L. Cheney. "Behavioural and Hormonal Responses to Predation in Female Chacma Baboons (*Papio hamadryas ursinus*)." *Proceedings of the Royal Society B* 273 (2006): 707–12. Quoted material, p. 709.

Fashing, Peter J., Nga Nguyen, Tyler S. Barry, C. Barret Goodale, Ryan J. Burke, Sorrel C. Z. Jones, Jeffrey T. Kerby, Laura M. Lee, Niina O. Nurmi, and Vivek V. Venkataraman. "Death among Geladas (*Theropithecus gelada*): A Broader Perspective on Mummified Infants and Primate Thanatology." *American Journal of Primatology* 73 (2011): 405–9. Quoted material, p. 408.

Mendoza, Sally, and William Mason. "Contrasting Responses to Intruders and to Involuntary Separation by Monogamous and Polygynous New World Monkeys." *Physiology and Behavior* 38 (1986): 795–801.

Sugiyama, Yukimaru, Hiroyuki Kurita, Takeshi Matsu, Satoshi Kimoto, and Tadatoshi Shimomura. "Carrying of Dead Infants by Japanese Macaque (*Macaca fuscata*) Mothers." *Anthropological Science* 117 (2009): 113–19.

VIDEO *Clever Monkeys*, narrated by David Attenborough (segment on toque monkeys starts at 1:15): http://www.youtube.com/watch?v=VaiFfSui4oc.

CHAPTER SEVEN

Anderson, James R. "A Primatological Perspective on Death." *American Journal of Primatology* 71 (2011): 1–5. Quoted material, p. 2.

Biro, Dora, Tatyana Humle, Kathelijne Koops, Claudia Sousa, Misato Hayashi, and Tetsuro Matsuzawa. "Chimpanzee Mothers at Bossou, Guinea Carry the Mummified Remains of Their Dead Infants." *Current Biology* 20 (2010): R351–R352. Quoted material, p. R351.

Boesch, Christophe, and Hedwige Boesch-Achermann. *The Chimpanzees of Tai Forest.* Oxford: Oxford University Press, 2000. Quoted material, pp. 248–49.

Goodall, Jane van Lawick. 1971. *In the Shadow of Man.* New York: Dell. Quoted material, p. 236.

———. *Through a Window.* New York: Mariner Books, 1990. Quoted material, pp. 196–97.

King, Barbara J. "Against Animal Natures: An Anthropologist's View." 2012. http://www.beinghuman.org/article/against-animal-natures-anthropologist's-view.

Sorenson, John. *Ape.* London: Reaktion Books, 2009. Quoted material, pp. 70, 85.

VIDEO Chimpanzee attack on Grapelli, narrated by David Watts ("Gang of Chimps Attack and Kill a Lone Chimp"; attack itself begins around 3 minutes in): http://www.youtube.com/watch?v=CPznMbNcfO8.

VIDEO Chimpanzee attack, narrated by David Attenborough: http://www.youtube.com/watch?v=a7XuXi3mqYM&feature=fvst.

CHAPTER EIGHT

Barash, David. "Deflating the Myth of Monogamy." *Chronicle of Higher Education*, April 21, 2001.

Heinrich, Bernd. *Mind of the Raven.* New York: Ecco, 1999.

———. *The Nesting Season: Cuckoos, Cuckolds, and the Invention of Monogamy.* Cambridge: Belknap Press, 2010. Quoted material, p. 26.

Marzluff, John M., and Tony Angell. *Gifts of the Crow: How Perception, Emotion, and Thought Allow Smart Birds to Behave Like Humans.* New York: Free Press, 2012. Quoted material, pp. 141, 146.

———. *In the Company of Crows and Ravens.* New Haven: Yale University Press, 2005. Quoted material, pp. 187, 195.

VIDEO The storks Rodan and Malena (narration in French): http://videos.tf1.fr/infos/2010/love-story-au-pays-des-cigognes-5786575.html.

CHAPTER NINE

ABC News. "Whales Mourn If a Family Member Is Taken: Scientists." August 20, 2008. http://www.abc.net.au/news/2008-08-10/whales-mourn-if-a-family-member-is-taken-scientists/470268.

Bearzi, Giovanni. "A Mother Bottlenose Dolphin Mourning Her Dead Newborn Calf in the Amvrakikos Gulf, Greece." Tethys Research Institute report (with photo). http://www.wdcs-de.org/docs/Bottlenose_Dolphin_mourning_dead_newborn_calf.pdf.

Evans, Karen, Margaret Morrice, Mark Hindell, and Deborah Thiele. "Three Mass

Whale Strandings of Sperm Whales (*Physeter macrocephalus*) in Southern Australian Waters." *Marine Mammal Science* 18 (2002): 622–43.

Klinkenborg, Verlyn. *Timothy, or Notes of an Abject Reptile*. New York: Vintage Books, 2007.

Ritter, Fabian. "Behavioral Responses of Rough-Toothed Dolphins to a Dead Newborn Calf." *Marine Mammal Science* 23(2007): 429–33. Quoted material, pp. 430, 431.

Rose, Anthony. "On Tortoises Monkeys & Men." In *Kinship with the Animals*, edited by Michael Tobias and Kate Solisti-Mattelon. Hillsboro, OR: Beyond Words Publishing, 1998. http://goldray.com/bushmeat/pdf/tortoisemonkeymen.pdf.

VIDEO Male sea turtle at memorial for Honey Girl: http://www.youtube.com/watch?v=qkVXucG1AeA.

VIDEO Dolphin-whale play: http://www.youtube.com/watch?v=lC3AkGSigrA.

VIDEO Still photographs and video related to whale mourning/whale strandings: http://www.youtube.com/watch?v=XaViQ7FHJPI.

CHAPTER TEN

Elephant Sanctuary. Account of Bella's death. http://www.elephants.com/elediary.php (begin at entry for October 24, 2011).

Holland, Jennifer. *Unlikely Friendships: 47 Remarkable Stories from the Animal Kingdom*. New York: Workman Publishing, 2011.

Pierce, Jessica. *The Last Walk: Reflection on Our Pets at the End of Their Lives*. Chicago: University of Chicago Press, 2012. Quoted material, pp. 220, 199.

Zimmer, Carl. "Friends with Benefits." *Time*, February 20, 2012, 34–39.

PHOTO Tarra and Bella together: http://www.elephants.com/Bella/Bella.php.

VIDEO Polar bears and dogs playing: http://www.dailymotion.com/video/x3ag90_polar-bears-and-dogs-playing_animal.

VIDEO *CBS Sunday Morning*, "The Common Bond of Animal Odd Couples": http://www.cbsnews.com/video/watch/?id=7362308n&tag=contentMain;contentBody.

PHOTO Tinky the cat at the piano: http://www.barbarajking.com/blog.htm?post=801721.

CHAPTER ELEVEN

ABC Science. "Lemmings Suicide Myth." April 27, 2004. http://www.abc.net.au/science/articles/2004/04/27/1081903.htm.

Bekoff, Marc. "Bear Kills Son and Herself on a Chinese Bear Farm." http://www.psychologytoday.com/blog/animal-emotions/201108/bear-kills-son-and-herself-chinese-bear-farm.

Birkett, Lucy, and Nicholas E. Newton-Fisher. "How Abnormal Is the Behaviour of Captive, Zoo-Living Chimpanzees?" *PLoS ONE* 6 (2011): e20101. doi: 10.1371/journal.pone.0020101.

Bradshaw, G. A., A. N. Schore, J. L. Brown, J. H. Poole, and C. J. Moss. "Elephant Breakdown." *Nature* 433 (2005): 807.

Guardian. "Dolphin Deaths: Expert Suggests 'Mass Suicide.'" June 11, 2008. http://www.guardian.co.uk/environment/2008/jun/11/wildlife.conservation1.
Karmelek, Mary. "Was This Gazelle's Death an Accident or a Suicide?" http://blogs.scientificamerican.com/anecdotes-from-the-archive/2011/05/24/was-this-gazelles-death-an-accident-or-a-suicide/.
King, Barbara J. "When a Daughter Self-Harms." http://www.npr.org/blogs/13.7/2012/07/12/156550195/when-a-daughter-self-harms.
Poulsen, Else. 2009. *Smiling Bears: A Zookeeper Explores the Behavior and Emotional Life of Bears*. Vancouver: Greystone Books. Quoted material, pp. 208–9.

CHAPTER TWELVE

Anderson, James R., Alasdair Gillies, and Louse C. Lock. "Pan Thanatology." *Current Biology* 20 (2010): R349–R351. Quoted material, p. R350.
Goodall, Jane van Lawick. *In the Shadow of Man*. New York: Dell, 1971. Quoted material, p. xi.
Teleki, G. "Group Response to the Accidental Death of a Chimpanzee in Gombe National Park, Tanzania." *Folia primatologica* 20 (1973): 81–94. Quoted material, pp. 84, 85, 89, 92, 93.

CHAPTER THIRTEEN

Berger, Joel. *The Better to Eat You With: Fear in the Animal World*. Chicago: University of Chicago Press, 2008. Quoted material, p. 117.
Bradbury, Ray. *Dandelion Wine*. New York: Doubleday, 1957.
Desmond, Jane. "Animal Deaths and the Written Record of History: The Politics of Pet Obituaries." In *Making Animal Meaning*, edited by Georgina Montgomery and Linda Kaloff, 99–111. East Lansing: Michigan State University Press, 2012. Quoted material, pp. 99, 100, 103, 104.
Lott, Dale F. *American Bison: A Natural History*. Berkeley: University of California Press, 2002. Quoted material, p. 4.
Whittlesey, Lee H. *Death in Yellowstone: Accidents and Foolhardiness in the First National Park*. Lanham, MD: Roberts Rinehart, 1995. Quoted material, pp. 4, 30.
PHOTO Martha Mason in her iron lung: http://www.nytimes.com/2009/05/10/us/10mason.html.

CHAPTER FOURTEEN

Archer, John. *The Nature of Grief: The Evolution and Psychology of Reactions to Loss*. New York: Routledge, 1999.
Didion, Joan. *The Year of Magical Thinking*. New York: Knopf, 2005. Quoted material, p. 27.
Goldman, Francisco. *Say Her Name*. New York: Grove Press, 2011. Quoted material, pp. 43–44, 240–41.

Lewis, C. S. *A Grief Observed*. New York: HarperOne, 1961. Quoted material, pp. 6, 9–10, 18, 25, 54, 72.
Oates, Joyce Carol. *A Widow's Story*. New York: Ecco, 2011. Quoted material, pp. 105, 275.
Rosenblatt, Roger. *Kayak Morning*. New York: Ecco, 2012. Quoted material, p. 143.
———. *Making Toast*. New York: Ecco, 2010. Quoted material, pp. 32–33.
Saunders, Frances Stonor. "Too Much Grief." *Guardian*, August 19, 2011. http://www.guardian.co.uk/books/2011/aug/19/grief-memoir-oates-didion-orourke.
Volk, Tyler. *What Is Death? A Scientist Looks at the Life Cycle*. New York: John Wiley and Sons, 2002. Quoted material, pp. 84–85.
VIDEO Gombe chimpanzees at waterfall, narrated by Jane Goodall: http://www.janegoodall.org/chimp-central-waterfall-displays.

CHAPTER FIFTEEN

Bar-Yosef Mayer, Daniella, Bernard Vandermeersch, and Ofer Bar-Yosef. 2009. "Shells and Ochre in Middle Paleolithic Qafzeh Cave, Israel: Indications for Modern Behavior." *Journal of Human Evolution* 56 (2009): 307–14.
Formicola, V., and A. P. Buzhilova. "Double Child Burial from Sunghir (Russia): Pathology and Inferences for Upper Paleolithic Funerary Practices." *American Journal of Physical Anthropology* 124 (2004): 189–98. Quoted material, p. 189.
Goldman, Francisco. *Say Her Name*. New York: Grove Press, 2011. Quoted material, pp. 306, 313.
Henshilwood, C. S., F. d'Errico, K. L. van Niekerk, Y. Coquinot, Z. Jacobs, S.-E. Lauritzen, M. Menu, and R. Garcia-Moreno. "A 100,000-Year-Old Ochre-Processing Workshop at Blombos Cave, South Africa." *Science* 334 (2011): 219–22.
Volk, Tyler. *What Is Death? A Scientist Look at the Cycle of Life*. New York: John Wiley and Sons, 2002. Quoted material, p. 83.
VIDEO/PHOTO Amos, Jonathan. "Ancient 'Paint Factory' Unearthed." BBC News: http://www.bbc.co.uk/news/science-environment-15257259.

AFTERWORD

Archer, John. *The Nature of Grief: The Evolution and Psychology of Reactions to Loss*. New York: Routledge, 1999. Quoted material, p. 1.
Sullivan, Deirdre. "Always Go to the Funeral." http://thisibelieve.org/essay/8/.

INDEX

Page numbers in italics indicate photographs.